Climate Smart Agriculture and Food Security

NIPA® GENX ELECTRONIC RESOURCES & SOLUTIONS P. LTD.
New Delhi-110 034

About the Authors

Sri Bidyadhar Maharana, a Post Graduate in Agronomy from Odisha University of Agriculture & Technology (OUAT) has about 55 years of government service in Agriculture Department of Odisha and Consultancy experience. He has received professional trainings from CRRI, GAA, MANAGE, NAARM, IIM, Lucknow etc. He was a National Professional in a GOI-UNDP assisted Sub-programme "Strengthening NRM and Sustainable Livelihoods of Women" in Tribal Odisha for five years. He was a freelancer Consultant in many WB, IFAD, DFID, ADB, UNDP, UNICEF, CARE and IWMI assisted projects in the state of Odisha for about 15 years. As a Certified trainer by Department of Personnel and Training, Government of India trained many high-level and middle-level officers on NRM, Watershed management, Project Planning, Micro-planning, Participatory Rural Appraisal, Extension Management and different subject matters for about 20 years in IMAGE and ICAR Institutes. He was a National Level Monitor (NLM) for Ministry of Rural Development, Government of India for monitoring Rural Development Schemes in Jharkhand, WB and Nagaland for three years. He has provided consultancy to reputed NGOs like Agragamee, XIM-CENDERET, CYSD, PRAGATI, NIRMAN, IGGSS, OPSL, RICOR, LAVS etc. for many years till now. He has conducted several studies for APICOL, OPSL, RICOR, IWMI, CARE, India, SAMBODHI, OTELP, WORLP, NABARD etc. He has developed user manuals on Land and Water Management for OWDM, OTELP, Decentralised GP planning for UNICEF, Women and Child Friendly GP Plan for SIRD, RKVY manual for Government of Odisha and prepared many projects for government. He has written many technical book-lets for UNDP and Nutritional Garden manual for IGGSS. He was associated with IWMI, Srilanka for preparing a Climate Smart Village Plan for Odisha. He has participated in Doordarsan and AIR as a resource person for about 20 years. He is a member of Senior Agricultural Scientist Association, Odisha.

Dr Ballav Chandra Nayak, M.Sc.(Ag) Ph.D. was Professor of Agronomy in Odisha University of Agriculture & Technology (OUAT) from 1995 to 2003 with 24 years of teaching and research experience. He has guided 24 M.Sc.(Ag) and three Ph.D. students in Agronomy. After his retirement he was a State Consultant, National Food Security Mission for two years. He has authored 7 books and participated in about 80 radio and television talks and imparted training to farmers in about 300 trainings. He has 56 research publications and 55 extension articles to his credit. He has edited and reviewed 5 books. He has attended 30 National and International Symposia and served as a resource person of 28 different organisations. He is a life member of 3 organisations and annual member of 7 organisations.

Climate Smart Agriculture and Food Security

Bidyadhar Maharana
Ballav Chandra Nayak

NIPA® GENX ELECTRONIC RESOURCES & SOLUTIONS P. LTD.
New Delhi-110 034

NIPA® GENX ELECTRONIC
RESOURCES & SOLUTIONS P. LTD.
101,103, Vikas Surya Plaza, CU Block
L.S.C.Market, Pitam Pura, New Delhi-110 034
Ph : +91 11 27341616, 27341717, 27341718
E-mail: newindiapublishingagency@gmail.com
www:www.nipabooks.com

For customer assistance, please contact
Phone : + 91-11-27 34 17 17
Fax : + 91-11- 27 34 16 16
E-Mail: feedbacks@nipabooks.com

ISBN: 978-93-58879-35-3

Composed and Designed by NIPA®.

ଓଡ଼ିଶା କୃଷି ଓ ବୈଷୟିକ ବିଶ୍ୱବିଦ୍ୟାଳୟ, ଭୁବନେଶ୍ୱର

ODISHA UNIVERSITY OF AGRICULTURE AND TECHNOLOGY

BHUBANESHWAR-751 003

ପ୍ରଫେସର ପ୍ରଭାତ କୁମାର ରାଉଳ

କୁଳପତି

Prof. Pravat Kumar Roul

Vice-Chancellor

Dated the 22nd September, 2023

Foreword

Agriculture plays a pivotal role in eradicating hunger, poverty and malnutrition. Climate change, however, has become a potential barrier to agricultural growth, especially in rainfed areas. On the other hand, agriculture has been an important contributor to global warming due to release of greenhouse gases like methane and nitrous oxide. Reducing agricultural emissions can play a significant role in mitigation of climat change.

Under such situations, the components of Climate Smart Agriculture (CSA) help in achieving food security by combating climate change effects. There is no single technology or approach to realize the requirement of climate smart agriculture. Site-specific assessment is required to recognize suitable technologies and practices based on local situation. Several climate resilient technologies have been worked out by various scientists at different locations for adoption and mitigation of climate change. Such technologies have been compiled in the present book entitled "Climate Smart Agriculture and Food Security". This book deals with the issues relating to climate change and its impact on agricultural activities. Various practices have been suggested to address climate change issues. The chapters deal with climate factors influencing agricultural production with focus on the dimension of CSA and climate smart village.

Authors of the book; Dr. B.C. Nayak and Sri Bidyadhar Maharana have long experience in teaching/training and research relating to agricultural sciences. Their vast knowledge and wide experience have been reflected by accommodating various facets of Climate Smart Agriculture and Food Security.

I believe, this book will provide a good understanding about role of climate on agriculture and possible technologies to deal with climate change issues. 1 hope, it will be useful for students, researchers, academicians and policy makers

(P.K. Roul)

Visit us at www.ouat.nic.in
Tel :0674-2397700 (O) Fax-0674-2397780
E-mail:vc@ouat.nic.in/vcouat@gmail.com

Preface

Agriculture is the mainstay of Indian economy. However, the contribution of agriculture and allied sectors to gross domestic product (GDP) is decreasing over the years. As per estimate of Ministry of Statistics and Programme Implementation, the Gross Value Added (GVA) of agriculture and allied sectors has come down to 18.3% in 2022-23. According to crop simulation models, it is estimated that rice yield in the country in rain-fed situation will be reduced by 20% in 2050 and 47% in 2080, if adoption of adaptation measures are not taken against climate change. Thus the climate change in recent years has challenged the food production in the world. In spite of sufficient global food supplies nearly 800 million people are hungry due to their inability to buy sufficient food for food security.

Indian farmers are facing second generation problems such as decreasing factor productivity, loss of soil fertility and bio-diversity, development of problematic soils, deficiency in irrigation system, and land degradation. These problems are further accentuated due to climate change that causes global warming, frequent weather change and loss of crop yields.

India has now exceeded China in population and such anthropogenic pressure affects agriculture, due to shrinking of resources, decline of land: man ratio and impact of climate change. Food security cannot be ensured by increasing production only. All the four dimensions of food security- availability, access, absorption and stability- must be ensured for achieving food security. The impacts of climate change on agriculture are seen in terms of temperature rise, variability in weather and climate pattern, change in water availability, sea water rise, soil salinization etc. which cause loss of crop yield and instability in crop production. In order to address such a situation, climate smart agriculture has been suggested by the Food and Agriculture Organisation (FAO). Climate smart agriculture has three pillars, such as food security, adaptation and mitigation. As agriculture is an important source of greenhouse gas emissions like methane and nitrous oxide, climate smart agricultural technologies and practices have to be adopted to manage the adverse impacts of climate change. In order to mitigate the climate change the emissions of greenhouse gases has to be reduced substantially and the soil carbon sink has to be enhanced by carbon sequestration, reduced tillage, managing soil health and soil organic matter etc. Climate smart agriculture mostly deals with these strategies. Climate smart village concept has been adopted in many countries and many states of India which broadly covers weather smart, seeds and breed smart, water smart, carbon smart, energy smart, nutrition smart and knowledge smart. One has to have clear understanding on these approaches before practicing in the farm.

The present book has been written covering the general understanding of weather and climate, impact of weather and climate on agriculture, challenges of climate change on food security, management of climate change by climate smart agriculture which covers adaptation, mitigation strategies and adoption of climate resilient agricultural practices in different chapters. Chapters 1 to 7 deal with basics of climate and weather, climate of India and Odisha, climate and crop production, climate change and its impact on agriculture, challenges of climate change to food security and basics of climate smart agriculture. Strategies for climate smart agriculture, practices for climate smart agriculture and climate smart village concept have been described in detail in Chapter8, 9 and 10. Extension approach to climate change has been explained in Chapter-11.

Both the authors have put their efforts to bring the readability of the book and incorporate all the recent findings of research quoting the authors in the book. Authors acknowledge the writers and researchers of different resources of literature which may have been referred during the writing of this book. We feel that the book shall be useful to the students, researchers, police makers, planners, field officers, non-Government Organisations and scholars appearing in competitive examinations.

Bidyadhar Maharana

Ballava Chandra Nayak

Contents

Acronyms and Abbreviations

ACZ	:	Agro-climatic Zones
AFLOU	:	Agriculture, forestry and other land use
AWD	:	Alternate Wetting and Drying
BBF	:	Broad Bed and Furrow system
CA	:	Conservation Agriculture
CCAFS	:	Climate Change, Agriculture and Food Security
CFC	:	Chlorofluorocarbon
CH_4	:	Methane
CO	:	Carbon Monoxide
CO_2	:	Carbon Dioxide
CPR	:	Common Property Resources
CRA	:	Climate Resilient Agriculture
CRIDA	:	Central Research Institute on Dryland Agriculture
CSA	:	Climate Smart Agriculture
CSV	:	Climate-Smart Village
DSR	:	Dry-seeded rice
ETL	:	Economic Threshold Level
F2FE	:	Farmer-to-Farmer Extension
FAO	:	Food and Agriculture Organisation
FFS	:	Farmers Field School
FTS	:	Fertiliser Tree Systems
GHG	:	Greenhouse gas
GM	:	Genetically Modified
GMO	:	Genetically Modified Organisms
GOI	:	Government of India
GWP	:	Global Warming Potential
HCFC	:	Hydro-chlorofluorocarbons
HFC	:	Hydro-fluorocarbons
IARI	:	Indian Agricultural Research Institute
ICAR	:	Indian Council of Agricultural Research
ICT	:	Information and Communication Technology
IFS	:	Integrated Farming System

INM	:	Integrated Nutrient Management
IPCC	:	Intergovernmental Panel on Climate Change
IPM	:	Integrated Pest Management
IRRI	:	International Rice Research Institute
ITK	:	Indigenous Technical Knowledge
IWRM	:	Integrated Water Resource Management
KVK	:	Krishi Vigyan Kendras
LCC	:	Leaf Colour Chart (LCC)
LLL	:	Laser land levelling
MGNREGA	:	The Mahatma Gandhi National Rural Employment Guarantee Act
N_2O	:	Nitrous oxide
NAPCC	:	National Action Plan on Climate Change
NARP	:	National Agricultural Research Project
NBSS & LUP	:	National Bureau of Soil Survey and Land Use Planning
NICRA	:	National Innovations in Climate Resilient Agriculture
NPV	:	Nuclear Polyhedrosis Virus
PAR	:	Participatory Action Research
PET	:	Potential Evapo-transpiration
PMKSY	:	Pradhan Mantri Krishi Sinchayee Yojana
PRA	:	Participatory Rural Appraisal
PVA	:	Participatory Vulnerability Assessment
RAM	:	Relatively Available Moisture
RCT	:	Resource Conserving Technologies
SAARC	:	South Asian Association for Regional Cooperation
SAMETI	:	State Agricultural Management, Extension and Training Institute
SO_3	:	Sulphur trioxide
SOM	:	Soil Organic Matter
SRI	:	System of rice intensification
SSNM	:	Site-specific nutrient management
UNFCCC	:	United Nations Framework Convention on Climate Change
VCRMC	:	Village Climate Risk Management Committee
WMO	:	World Meteorological Organisation
WUA	:	Water Users Association

1

Weather and Climate

1. Weather and Climate

Weather and climate have profound influence on agriculture of a region. There are some differences between these two terms. Weather describes the state of the lower atmosphere of a place with respect to temperature, precipitation, wind direction, wind speed, humidity, cloudiness and pressure etc. over a short period of time. The short term variations over minutes to days are indicated as weather of a location. It says whether the atmosphere is hot or cold, rainy or dry, sunny or cloudy, still or cyclonic and so on usually for period of 24 hours. The weather may vary from place to place and from time to time due to difference in temperature, cloudiness, humidity, rain and other events of atmosphere. It may be sunny at one locality, but rainy at another locality. Similarly the weather of a place in winter morning may be foggy and sunny after a few hours as the fog disappears. Variation in weather is driven by Sun as it heats air in the lower atmosphere at varying degrees. Meteorology is the study of weather and the meteorologists forecast the weather of a station basing on the observations recorded in the station or date collected from satellites Weather prediction is done by analysing the meteorological data from satellites and weather stations, which is very important as weather influences many human activities. Important activities influenced by weather are agriculture, sailing, fishing, air flight, fire fight, trading, and military operations etc.

Climate is defined as statistical weather information that indicates the variation of weather at a given place for a specific interval. Climate usually describes the atmospheric changes over longer periods of time, usually for 30 years or more as defined by World Meteorological Organisation (WMO). Usually climate is understood as the long-term weather pattern pattern in a particular area. The atmospheric events like cold spell or dry spell indicates the weather of the place and not the climate. Variable climates are seen in different parts of the world. It is extremely cold at the poles, hot and rainy at the equator, temperate at mid-latitudes, wet with cold winter and hot summer in east coast continents, dry in interior parts of continents, and mild with rainy winter and dry summer in continental west coasts. Generally Earth's climates are classified as tropical, dry, temperate, continental and polar.

Both weather and climate have some common components like precipitation, wind speed and direction, humidity, atmospheric temperature and pressure. Climate is considered as average weather conditions of a region over time. A change in climate may lead to change in weather patterns of a place. Weather may change over a period of time that may be hour to hour, day to day and season to season. However, climate is the average of weather over time and space.

Differences between weather and climate

Weather and climate are both used interchangeably sometimes, but differ in terms of the length of time they measure and what trends affect them. The climate and weather is usefully summarized by the popular phrase "Climate is what you expect, weather is what you get." The scientists describe climate as averages of precipitation, temperature, humidity, sunshine, wind velocity, phenomena such as fog, frost, and hail storms, and other measures of the weather over a long period of 30 years.

Table 1.1: Difference between climate and weather

	Climate	Weather
Definition	Climate refers to statistical weather information that indicates the variation of weather at a given place for a specified interval.	Weather is the day-to-day state of the atmospheric situation and its short-term variation.
Components	Climate includes precipitation, temperature, humidity, sunshine, wind velocity, fog, frost, and hail storms over a long period of time	Weather includes temperature, sunshine, rain, cloud cover, winds, hail, snow, sleet, freezing rain, flooding, ice storms, thunderstorms, heat waves etc. as short term events
Forecast & time	By aggregates of weather statistics over periods of 30 years	By collecting meteorological data, like air temperature, pressure, humidity, solar radiation, wind speeds and direction etc. over a short period of time, usually 24 hours.
Study	Climatology	Meteorology

In order to understand the climate change, climate study is important. Consistent rise in global temperature is expected to raise sea levels, and change precipitation pattern and other local climate conditions. Climate change in a region may alter forests, crop yields, and affect human health, animals, and many types of ecosystems.

1.1 Elements of Climate & Weather

Although various elements make up the climate of a region, temperature, atmospheric pressure, wind speed and direction, humidity, precipitation, visibility, cloudiness, solar radiation are the most common. These are the primary elements of weather and climate.

Temperature: Temperature is the amount of heat energy experienced in the air. It is measured in Celsius degrees or Fahrenheit degrees. Sun is the primary source of energy which heats up the atmosphere and Earth due to its incoming radiation. The solar radiation incepted by the Earth's surface is nearly two calories per square centimetre per minute which is equivalent to 2 Langley. Earth's radiation balance is maintained between incoming solar radiation and outgoing radiation. Nearly one-half of heat energy radiated as sun light and is deflected back by clouds, water vapour and atmospheric dust into space. The remaining heat is absorbed by the soil and water. Temperature variation during day and seasons is caused due Earth's rotation and translational motion of the Earth around the Sun. Temperature is very important factor in determining the weather as it controls other elements , such as precipitation, humidity, clouds and atmospheric pressure.

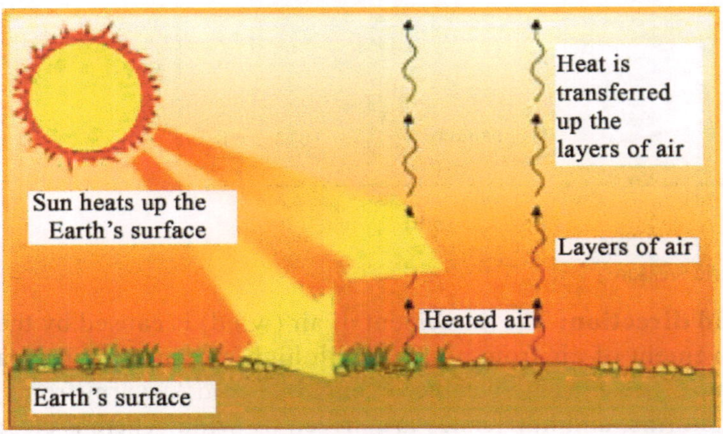

Fig. 1.1: Heating effect of Sun

Atmospheric pressure: Atmospheric pressure is created by weight of air on a given surface of Earth's atmosphere. It is the total weight of a mass of column of air per unit area at sea level. The air is a mixture of several gasses like nitrogen, oxygen, argon, carbon dioxide and a few other gases which have certain weight. Usually the atmospheric pressure is the maximum at sea level due to presence of more air close to ground and it goes on decreasing as the altitude increases. It varies vertically and the values decrease as we ascend in altitude. As elevation increases, there is less overlying atmospheric mass, which causes decreased pressure with increasing elevation. Usually atmospheric pressure decreases with increasing altitudes at a rate of 0.1 inch or 3.4 millibar (mb) per 600 feet. The atmospheric pressure is also called barometric pressure, named after the instrument used to measure air pressure. The atmospheric pressure is usually measured by a barometer, which is 760 mm or 1013.2 mb at sea level.

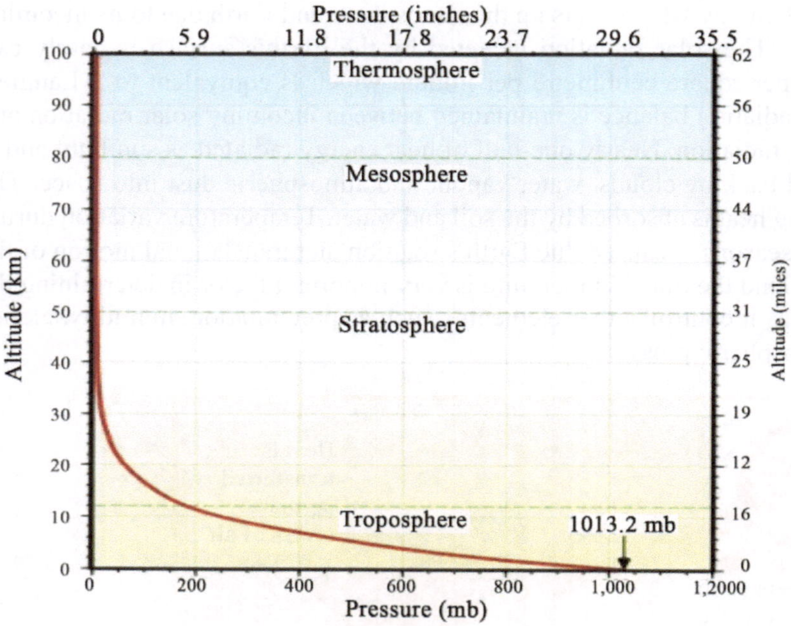

Fig. 1.2: Atmospheric pressure

Wind speed and direction: The movement of air (wind) is caused by the large-scale movement of air from an area with high pressure to an area with low pressure. The extreme weather events like cold & warm fronts, clouds, thunderstorms, and hurricanes are driven by wind. There is close relationship between pressure gradient and circulation of air. Control of air motion is regulated by the pressure gradient, Earth's rotation, and coriolis force, frictional force, and centrifugal action of wind etc. The force which deflects the direction of winds is called 'Coriolis Force'. The Earth rotates from west to east and completes one rotation in 24 hours.

The wind systems in atmosphere caused by solar radiation and Earth are called planetary wind. Such planetary winds effect climate as they cause the land to cool down. They also cause atmospheric moisture to condense and form clouds. Planetary winds are classified as Trade winds or Tropical Easterlies, Westerlies and Polar Easterlies. Wind speed and its strength are determined by the distance between the low pressure and high-pressure areas and the difference in air pressure. Wind speed is measured by Anemometer and the direction is measured by a Wind vane or Wind cock.

The winds that change their direction with onsets of different seasons are called as seasonal winds or monsoon winds Monsoons were traditionally explained as "land and sea breezes on a large scale". The monsoon drives vital seasonal rainstorms that provide water for crops and forests. Excess of monsoon causes floods while deficit monsoon causes drought. The monsoon could be seen in summer or winter season (Fig.1.3).

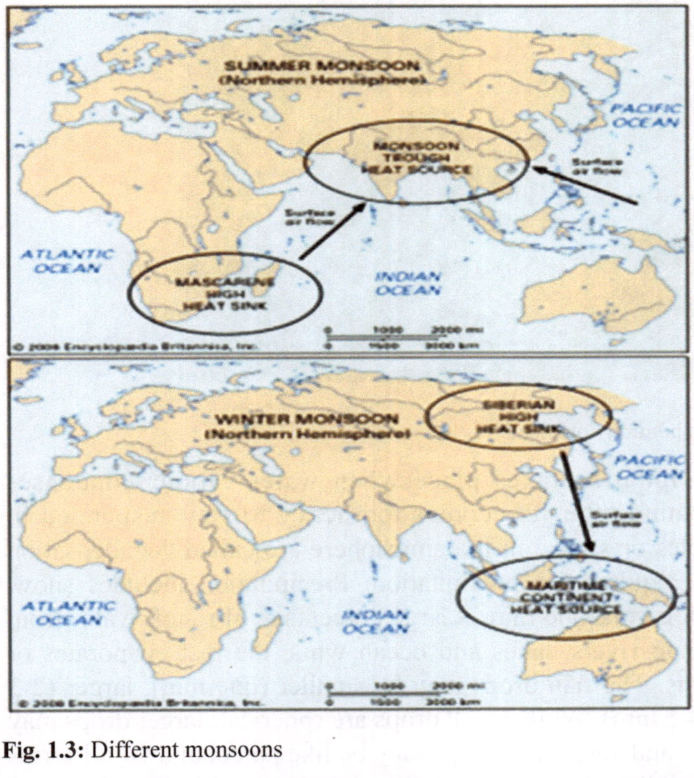

Fig. 1.3: Different monsoons

Humidity: The amount of water vapour present in the atmosphere at a given time is called humidity. Since water vapour is formed after evaporation of water, it is not visible. Humidity plays a big part in weather formation and directly influences our physical comfort levels. Humidity is measured by an apparatus called Hygrometer and is expressed as absolute humidity, relative humidity and specific humidity. Absolute humidity is the amount of water vapour in atmosphere per unit volume of air at a definite temperature. The ratio of water vapours present in air having a definite volume at a specific temperature compared to the maximum water vapours that the air can hold at that given temperature is called relative humidity (RH). Specific humidity is the mass of water vapour in grams contained in a kilogram of air which represents the actual quantity of moisture present in a definite volume of air. Humidity is very important for any ecosystem which can influence many functions. If saturated air is warmed, it can hold more water vapours. Cloud is formed when invisible water vapour in atmosphere condenses into visible water droplets and eventually precipitates.

Fig. 1.4: Effects of high humidity on plants

Precipitation: Precipitation takes place when water vapour condenses and falls to the ground after it becomes too heavy to stay suspended in the air. Dust particles or smoke in the atmosphere are called "condensation nuclei," which are required for precipitation. Precipitation includes snow, hail, sleet, drizzle, fog, mist and rain. A large percentage of runoff water from watersheds drains into rivers, lakes and ocean while the rest evaporates or passes through plants. The rain drops may be smaller (one mm), larger (2-3 mm)or very large (4.5 mm). While small drops are spherical, larger drops may be like kidney beans and very large drops may be like parachute. Of all forms of precipitation, rainfall is a major source which is measured in a rain gauge installed in rain recording station. Precipitation is always considered as fresh water, even when the water originated from the ocean. However, pollutants in the atmosphere may contaminate water droplets before they fall to the ground. Acid rain is often is caused due to such pollutants. The rainfall pattern involving quantity, quality and distribution are very important for crops grown in rain-fed areas.

Fig. 1.5: Rainfall

Visibility: Visibility is a critical element of weather which measures the degree through which an object can be observed from a certain distance. Visibility is obstructed by cloud, mist, haze, fog, and precipitation. Visibility plays a crucial role at airports and harbours. It is represented by the meteorological optical range (MOR) and sensors like "forward scatter sensor" are used to measure visibility. In extremely clean air the visibility can be up to 70 kilometres (43 mi) to 100 kilometres (62 mi). Due to air pollution and high humidity visibility is reduced. In sandstorm in desert areas and forest fires, visibility is often reduced. Blizzards (blowing snow) are also defined in part by low visibility.

Cloudiness: Clouds are suspended water molecules in different states that are formed in atmosphere when humid air cools down to its dew point, and water droplets or ice attach to small particles of dust, ash, or other contaminants. Clouds are indicators of probability of rain. Clouds move in one direction other than wind and indicate the occurrence of rain. Atmospheric pressure and humidity affect the shape, size, and type of clouds. The typology of clouds or classification was proposed by Luke Howard as low clouds, average clouds and high clouds.

Low clouds: Stratus, nimbostratus, stratocumulus, cumulus, towering cumulus and cumulonimbus

Average clouds (6500-20000ft): Altostratus, altocumulus and lenticular.

High clouds (above 20000 ft): Cirrus, cirrocumulus and cirrostratus.

These clouds appear in different forms such as , (i)Cirro-: curl of hair, high; (ii)Alto-: mid; (iii) Strato-: layer; (iv) Nimbo-: rain, precipitation; and (v) Cumulo-: heap.

Clouds play multiple critical roles in the climate system. Satellite and radar images are used to measure cloud density, the amount of moisture, the temperature, and movement of the clouds.

Clouds are the base for precipitation and they efficiently reflect light to space and thus contribute to the cooling of atmosphere. During summer clouds provide protection from sun rays and in winter reduce nocturnal radiation and check the fall of temperature. A small increase in cloud cover can balance the heating resulting from greenhouse gases.

Fig.1.6: Clouds in sky

Solar radiation: Sunshine duration is the length of time the Earth's surface is directly exposed to solar radiation. It is also referred to bright sunshine hours (BSH).The sunshine caused by solar radiation greatly influences other elements of the weather like ambient temperature, humidity and air pressure, which can change the whole makeup of the weather conditions. Sunshine recorders, more specifically Campbell–Stokes recorders, are the instruments used to record sunshine duration.

Solar radiation heats the Earth's surface which in turn raises the air temperature. It drives evaporation, so long as there is water. Heating of Earth's surface creates pressure gradients that result in wind. Solar radiation that reaches the Earth's surface varies according to geographic location, time of day, season, local landscape and weather. Solar radiation comes in many forms, such as visible light, radio waves, heat (infrared), x-rays, and ultraviolet rays. Solar radiation is higher on clear, sunny day and usually low on cloudy days.

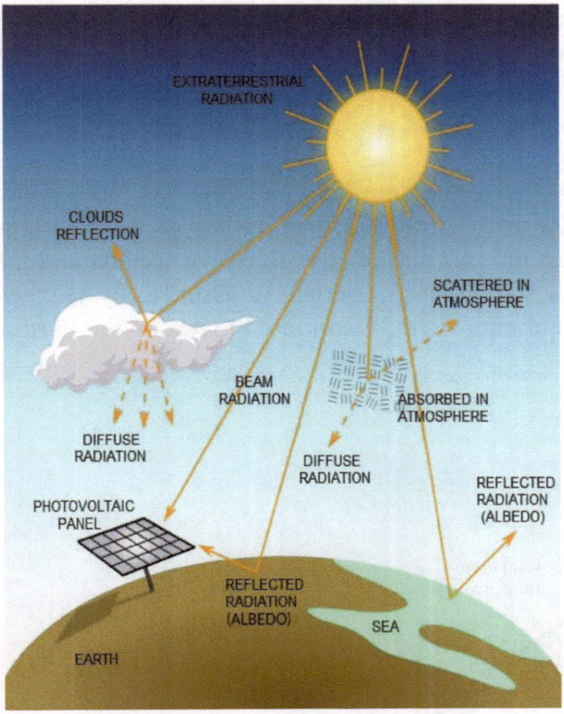

Fig.1.7: Solar radiation

1.2 Climate and Season

While climate is the average weather conditions in a certain area over a long period of time, the seasons are the changes in weather that occur throughout the year. The seasons are the changes in the climate throughout the year in terms of different temperatures, sunlight and precipitation. Considering a number of factors like latitude, altitude, and proximity to water, the climate of a place can be called as temperate, tropical, or polar. Temperate climate is characterised by warm summer and cool winter whereas tropical climate has hot summer and warm winter. Polar climate is experienced with cold winter and cool summer. Seasons are caused by the Earth's tilt on its axis as the sun's rays hit the Earth at different angles throughout the year. The main seasons are spring, summer, fall, and winter. In India the main seasons are rainy (wet), winter and summer (dry) with intermediary seasons like autumn, pre-winter and spring.

1.3 Climate Classification

Climate classification has been made for the scientific understanding of climates of different regions. Modern classification of climates has been started from the 19th century using empirical and generic methods. But two popular methods, such as Koppen and Thornthwaite methods are invariably used for classification of climate.

1.1.1 Köppen Classification System

Wladimir Köppen (1846-1940) gave the empirical classification and several revised versions thereafter. He used certain critical values of temperatures of the warmest and the coldest months and of rainfall of the wettest and the driest months coinciding with vegetation divisions. As assumed by Koppen vegetation was used to represent major ecosystems and biomes on the Earth.

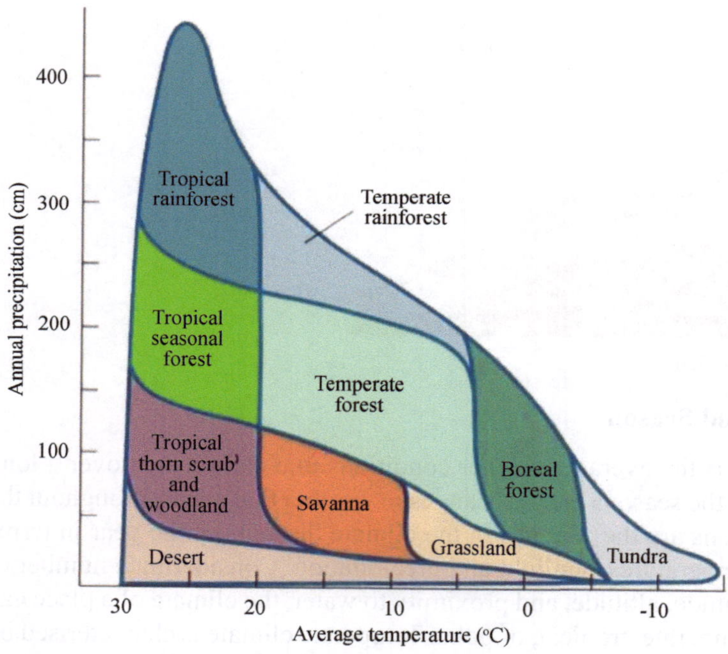

Fig. 1.8: Temperature and precipitations vs vegetation

Koppen considered mean monthly and annual temperature and mean monthly and annual precipitation for delineation of climatic regions. He classified the climate of world into five climate zones based on temperature, which allowed different vegetation growth. Köppen's map showed different colours and shades to represent the different climate zones of the world.

Zone A: covers tropical or equatorial zone shown in blue colour

Zone B: covers arid or dry zone indicated by red, pink, and orange colours

Zone C: covers warm/mild temperate zone indicated in green colour

Zone D: represents continental zone indicated in purple, violet, and light blue colours

Zone E: covers polar zone indicated by grey colour

These zones are expressed as tropical, dry, mild, cold temperate and polar.

Tropical (A): In tropical climate the temperature in cooler months does not exceed 18^{0}C. Annual precipitation is more than 1500 mm.

Dry Climates (B): In this zone the precipitation is little and no month is more than average temperature of 10^{0}C . Constant water deficiency is experienced.

Humid mesothermal(C): Warm and humid summer with mean temperature of coldest month is below 18°C but above -3°C and that of the warmest month is above 10°C.

Humid micro-thermal or cold forest climates(D): Warm to cool summer with the coldest month temperature below -3^{0}C and warmest month temperature above 10^{0}C.

Polar climates(E): Year-round cold with temperature of every month is below 10°C and without warm season.

Koppen used English capital letters (A, B, C, D, E, F, M, S, T, W) and small letters (a, b, c, d, f, h, k, m s, w) for climatic classification. The capital letters indicated the characteristic of vegetation and small letters indicated the precipitation and heat. Koppen's scheme uses letter symbols to denote various characteristics, which is practical and convenient. However, it ignores other factors, such as cloudiness, wind, rainfall intensity, currents and, above all, the air masses which form the basis of modern climatology. It is also difficult to explain the existence of different vegetation types within the same climatic division and similar vegetation types in different climatic divisions.

The sub-divisions with characteristics of these five groups are described in Table 1. 2.

Table 1.2: Climatic classifications by Koppen

Group	Type	Region	Characteristics
A-Tropical	Af = Tropical rainforest (equatorial)	Amazon basin, Zaire basin and south-east Asia, towns such as Kishangani (Congo), Singapore and Hawai (USA) ,	Characterized by temperature exceeding 18^{0}C and mean annual temperature of 20^0 C, rainfall with annual average of 250 cm suitable for luxuriant vegetation..
	Am= Tropical monsoon	Abidjan (Ivory Coast), Jakarta (Indonesia), Yangon (Myanmar) and Chittagong (Bangladesh), Pacific coast of Colombia, Guinea coast of west Africa, south-east Africa, south and south-east Asia and northern Australia	Seasonal reversal of winds, associated with alternate periods of rainfall and drought with a short dry season. Less than 60 mm precipitation in driest month. Rainfall occurs excessively during summer rainy season making land surface moist throughout the year. It is typical of monsoon climate.
	As = Dry summer	Chennai (India) and Odisha is a typical example of summer dry (As)	Rain-shadow areas remain dry during summer monsoon and receive winter rainfall from retreating monsoons
	Aw=Tropical savanna	Mumbai (India), Dhaka (Bangladesh), Dares Salam (Tanzania), Darwin (Australia) and Ho Chi Minh City (Vietnam)	Winter season is usually dry with annual mean temperature of 23^{0}C and rainfall is 160 cm. The water balance remains in deficit during dry season and surplus during wet season.
B – Dry (arid and semiarid)	BWh =Arid climates (desert)	Khartoum (Sudan), Doha (Qatar), and Death Valley (USA) west Asia, Thar desert, and central Australia	Low altitude sub-tropical high pressure region with mean annual temperature of 38°C and scanty and erratic rainfall. Vegetation varies with the soil type.
	BWk= Mid-Latitude deserts	Turpan (China), and Nukush (Uzbekistan) are typical examples of the climate.	These are mid latitude cold deserts. Atmospheric conditions are cold and dry . Annual average temperature is 21^{0}C.
	BSh: Hot semi-arid climates	Lahore (Pakistan), Odessa (USA) and Niamey (Nigeria)	These are semiarid tropical steppe climates where average annual temperature is more than 18^0 C. Rainfall is meagre 30 cm per annum.
	BSk=Cold semi-arid climate	Baku (Azerbaijan), Denver (USA), Lhasa (China) and Ulaanbaatar (Mongolia)	These are temperate steppe climates with average annual temperature below 18^{0}C. These climates are usually found in the proximity of cold deserts.

Group	Type	Region	Characteristics
C – Mild humid mesothermal (Temperate climate)	Cfa: Subtropical wet climates	Within 25° to 45° latitude on east coast in both hemispheres e.g., south-east USA, southern Brazil, Uruguay, Argentina, and south-eastern Africa, eastern coastal belt of Australia, eastern China and Japan.	Hot and humid summer and mild winter with average annual temperature is 20°C and well distributed rainfall (100 cm).
	Cfb: Temperate marine climate	Paris (France), Amsterdam (Netherlands), Port Elizabeth (South Africa) and Melbourne (Australia) are examples of the climate	Short cool summers, mild winters with average annual temperature around 10°C with adequate precipitation in all the seasons and rainfall is 140 cm.
	Cfc: Sub polar marine climate	Unalaska (USA), Punta Arenas (Chile), and Reykjavik (Iceland) are examples of the climate.	Relatively cold in winter season. Summer season is short and the average temperature for a minimum of four months is around 10^{0}C. No significant precipitation difference between seasons
	Csa: Hot summer Mediterranean climate	Madrid (Spain), Dushanbe (Tajikistan), Perth (Australia) and Beirut (Lebanon) .	Coldest month averaging above 0 °C or −3 °C , at least one month's average temperature above 22 °C , and at least four months average above 10 °C . Summer season is hot and dry with less than 40 mm rain but winter season is normal and wet.
	Csb: Warm summer Mediterranean climate	Santiago (Chile), Cape Town (South Africa) and San Francisco (USA) .	It has a short summer season which is dry and hot. The winter season is usually favourable as it receives rainfall. Coldest month averaging above 0 °C or −3 °C , all months with average temperatures below 22 °C , and at least four months averaging above 10 °C . The driest month of summer receives less than 40 mm
	Csc: Cold summer Mediterranean climate	Rare instances of this climate can be found in some coastal locations in the North Atlantic and at high altitudes in Hawaii.	Coldest month with average temperature 0 °C −3 °C and 1–3 months averaging above 10 °C . Driest month of summer receives less than 40 mm of precipitation.

Group	Type	Region	Characteristics
	Cwa: Subtropical monsoonal climate	Islamabad (Pakistan), New Delhi (India), Kathmandu (Nepal) and Hong Kong .Southern Cone of South America, the Gangetic Plain of South Asia, south-eastern Africa, and parts of East Asia and Mexico.	Coldest month with at least one month's average temperature above 22 °C and at least four months averaging above 10 °C . 70% or more of average annual precipitation is received in the warmest six months. Winter season is normal and dry.
	Cwb: Highland type	Central America, South America, Africa, and South and Southeast Asia	Highland type with temperature remaining less than 22^0 C in hottest month. Winter season is cold and dry while summer season is warm and wet.
	Cwc = Cold subtropical highland climate	Rarely found mainly in isolated locations of Andes in Bolivia and Peru, as well as in sparse mountain locations in Southeast Asia.	Coldest month with average temperature above 0 °C or −3 °C and average temperature of 10^0C in 1-3 months. 70% or more of average annual precipitation is received in the warmest six months.
D-Cold temperate (humid micro thermal)	Dfa: Hot summer humid continental climate	New York city (USA), Almaty (Kazakhstan), Beijing (China) and Seoul (South Korea) .	The annual range of temperature is high and weather is highly variable. Summers are warm and winters are cold. Precipitation occurs in all the months.
	Dfb: Warm summer humid continental climate	Found in eastern Asia, China, Korea, and Japan generally found between 45^0 to 60^0 North latitudes	More rainfall occurs in summer season and there is a drier winter.
	Dwa: Monsoon influenced hot- summer humid continental climate	It is commonly found in north eastern U.S.A. and south Eastern Canada. Pyongyang (North Korea) and Tianjin (China) are the typical examples of the climate.	Both coniferous and deciduous forests are found in this belt. The climate is characterized by moist continental with short period of summer dryness.
	Dwb: Monsoon-influenced warm-summer humid continental climate	Found along oceanic margins of the continents in higher latitudes. Vladivostok (Russia), and Pembina (USA) are representative stations of the climate	It is wet climate with long dry and severe cold winter season, short and cold summer season.
	Dwc: Monsoon-influenced subarctic climate	Moron (Mongolia) and Yushu City (China) are examples of the climate	It is a sub- arctic climate characterized by long cold and dry winter season. Summer season is short and mild

Group	Type	Region	Characteristics
	Dwd: Monsoon-influenced extremely cold subarctic climate	Found in the proximity of Sub-Arctic region. Yakutsk (Russia) and Oymyakon (Russia) .	It is characterized by long dry and extreme cold winter season. Summer season is short and mild warm.
	Dsa: Mediterranean influenced hot-summer humid continental climat	Sanandaj (Kurdistan-Iran), Hakkari (Turkey) are examples of the climate.	Average temperature of the coldest month is below 0°C or -3 °C while average temperature of the warmest month is below 22 °C. Average temperature at least four months is above 10 °C. The driest month of summer receives less than 30 mm of precipitation.
	Dsb: Mediterranean-influenced warm-summer humid continental climate	Sivas (Turkey), Roghun (Tajikistan), Dras (India) and Bridgeport (CaliforniaUSA) .	Average temperature of the coldest month below 0 °C or −3 °C while average temperature of the warmest month is below 22 °C. Average temperature of at last four months is above 10 °C . The driest month of summer receives less than 30 mm precipitation.
	Dsc: Mediterranean-influenced subarctic climate	Homer (Alaska, USA), Mount Olympus (Greece) and Handud, (Badkhshan Province, Afghanistan) .	Temperature recorded in the coldest month averaging < 0 ° C and 1-3 months averaging >10° C. There is at least three times as much precipitation in wettest month of winter as in driest month of summer.
	Dsd: Mediterranean-influenced extremely cold subarctic climate	Found in eastern Siberia, Seymachan, (Magadan Oblast, Russia)	Average temperature of coldest month is below −38 °C and 1–3 months averaging above 10 °C .The driest month of summer receives less than 30 mm of precipitation.
E – Polar	ET: Tundra type of climate	Stanley (Falkland Islands), Mount Wellington (Australia) and La Rinkonada (Peru)	It is characterised by a short summer season .Average temperature of warmest month lies between 0 °C and 10 °C.
	EF: Polar climate (ice cap climate)	Dominant in Antarctica, inner Greenland, and summits of many high mountains, even at lower latitudes Mount Fuji (Japan	Characterised by eternal winter with average temperatures below 0 °C .

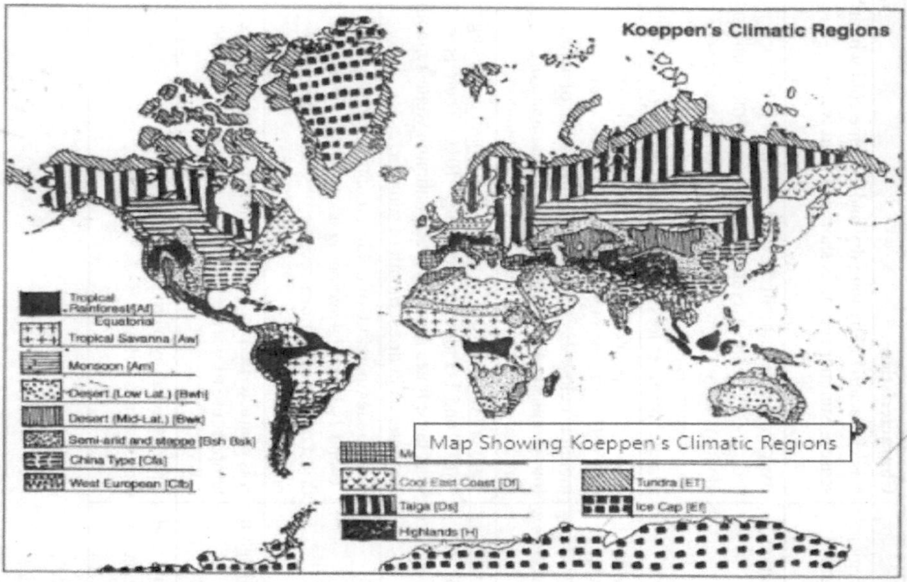

Fig. 1.9: Koepen's climatic regions

1.1.2 **Thornthwaite Classification**

Thornthwaite classified climates based on moisture and heat as primary factors and seasonality as secondary factors in 1931 and further modified in 1945. It was an improved and modified version of Koeppen's classification. Thorthwaite assumed that there was correlation between precipitation, temperature and plant association. The growth and development of natural vegetation depends on precipitation, rate of evaporation and thermal conditions of a region. The classification of climates was based on the concept of potential evapotranspiration (PET), which is equal to the total amount of water lost by way of evaporation and transpiration over an area under sufficient moisture condition. While classifying the climates, Thornthwaite used several factors.

(i) Moisture factor

According to Thornthwaite the precipitation (P) alone is not a good indicator of moisture conditions for which he developed the concept of potential evapotranspiration (PET) calculated from the temperature and day length. With the use of PE in combination of evaporation, water budget methodology was used to develop a moisture index. Moisture index was calculated from aridity and humidity indices.

Aridity Index (*Ia*)= (D/PET).100

Humidity Index (*Im*)=(S/PET).100

Where, *D* is the annual water deficit, *S* is the annual water surplus, and *PET* is the annual potential evaporation.

Moisture Index is given by:

$I_m = I_h - 0.6xI_a = (100s-60d)/n$

Using moisture index (I_m) climate of world was grouped as moist and dry (Table 1.3)

Table 1.3: Climate types and moisture index in Thornthwaite's classification

Symbol	Climate type	Moisture Index	Broad group
A	Per humid	100 and above	Moist climate
B4	Humid	80 to 100	
B3	Humid	60-80	
B2	Humid	40-60	
B1	Humid	20-40	
C2	Moist Sub-humid	0-20	
C1	Dry Sub-humid	-20 to 0	Dry climate
D	Semi-arid	-40 to -20	
E	Arid	-60 to -40	

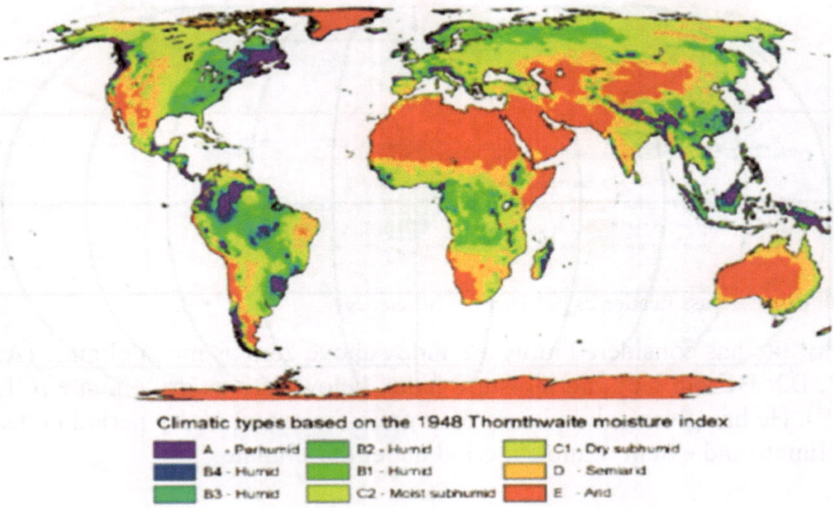

Source:https://www.researchgate.net/publication/250171991_A_Revised_Thornthwaite-Type_Global_Climate_Classification

Fig.1.10: Climate classification based on moisture index

(ii) Thermal factor

Thornthwaite used thermal factor and called Index of Thermal Efficiency (TE) on PE as he considered both TE and PE are important factors for productivity of environment. The climate classification as per TE index was as follows:

Table 1.4: Thornthwaite climate type as per Thermal Index

Climate type	TE index x PE; cm
A' Megathermal	114 and above
B'4 Mesothermal	99.7-114
B'3 Mesothermal	60-80
B'2 Mesothermal	40-60
B'1 Mesothermal	57-71.2
C'2 Microthermal	0-20
C'1 Microthermal	-20 to 0
D'Tundra	-40 to -20
E' Frost	-60 to -40

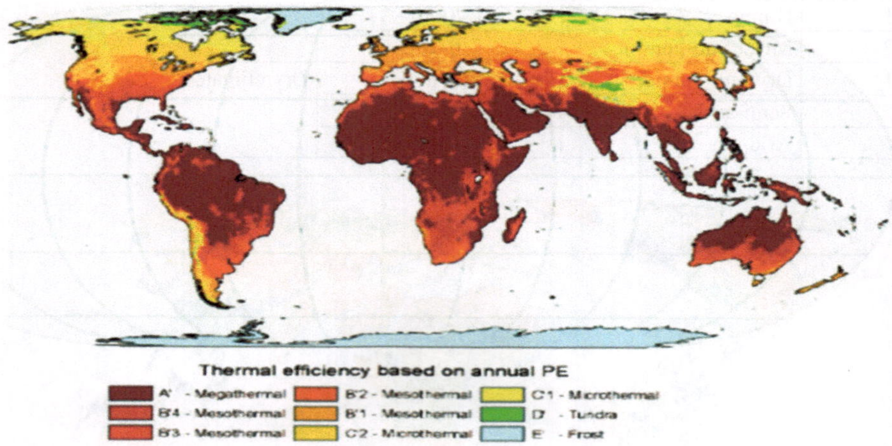

Fig. 1.10: Climate classification as per Thermal Efficiency

Thornthwaite has considered moisture index above zero as moist climate (A, B4, B3, B2, B1 and C2) and moisture index below zero as dry climate (C1, D and E). He has given subdivisions to express the extent of dry period under moist climate and extent of moist period under dry climate.

Table 1.5: Subdivisions of moist and dry climates in Thornthwaite's classification

Moist-climate description	Aridity index	Dry-climate description	Humidity index
R- Little or no water deficiency	0-10	D- Little or no water surplus	0-16.7
S_1- Moderate summer water deficiency	10-20	S_1-Moderate summer water surplus	16.7-33.3
W_1- Moderate winter water deficiency	10-20	W_1-Moderate winter water surplus	16.7-33.3
S_2- Large summer water deficiency	≥ 20	S_2-Large summer water surplus	≥33.3
W_2-Large winter water deficiency	≥20	W_2-Large winter water surplus	≥33.3

1.4 Factors influencing the World Climatic Regions

Among the most important factors that affect the world climate the most important natural factors are: latitude, altitude, wind and air mass, distance from the sea, ocean currents, direction of prevailing wind, land topography, the El Nino phenomenon and human effects.

1.4.1 Latitude (Distance from the equator): Due to curvature of Earth, the temperatures drop in areas far from the equator. This is because the Sun's rays are dispersed over a larger area of land as we move away from the equator. This is the most important, and it is based on the concentration of sunlight and the area that it affects. In areas closer to the poles temperatures are cooler. In addition, the presence of ice and snow nearer the poles cause a higher albedo that reflects the solar energy for which it gives a cooling effect.

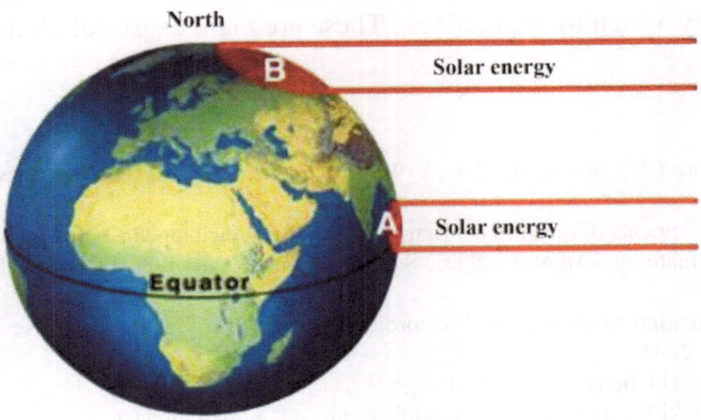

Fig. 1.10: Effect of Latitude on climate

1.4.2 Altitude: The higher altitudes usually have colder temperatures areas. Generally temperature decreases by 1°C for every 100 metres in altitude. Therefore hilly areas at higher altitudes have cooler climate than the plain areas

1.4.3 Wind and air mass: Heated ground causes air to rise which results in lower air pressure. When air rises it causes cooling and as it descends to the ground it causes high air pressure. These air masses absorb the climate of the air below it.

1.4.4 Distance from the sea (Continentality): Oceans heat up and cool down much slowly than the inland areas. This is the reason why coastal locations tend to be cooler in summer and warmer in winter than inland areas at the same latitude and altitude.

1.4.5 Ocean currents: The rotation of earth affects the oceans setting up currents that flow within the ocean basins. Ocean currents are driven by surface winds, Earth's rotation, and differences in salinity. Under the influence of the trade winds, surface waters near the equator flow from east to west. Warm water pools along the west coast of continents, which sets up a temperature gradient across the ocean surface. Disruption of this temperature gradient creates the event known as El Niño. El Niño refers to the irregular warming of surface water in the Pacific.

1.4.6 Human influence: Apart from natural factors, the human factors are also responsible to affect the climate. With the rise in human population, deforestation has increased the carbon dioxide concentration in the atmosphere. Emission of greenhouse gases (GHG) such as carbon dioxide, methane, nitrous oxide, etc. increases due to consumption of fossil fuel to run the vehicles and machinery and faulty agricultural practices . These are main causes of climate change.

References

Environment and Climate Change Canada (2015). Weather elements. Retrieved from: www.gc.ca.

Geoencicopadia (s / f). Elements of the climate. Retrieved from: geoenciclopedia.com.

Hindore, J.H.(2009), Climatology-An Atmospheric Science, Prentice Hall of India Learning Pvt Ltd, New Delhi.

Holton, J. R. An Introduction to Dynamical Meteorology, 4th ed. Burlington, MA: Elsevier Academic Press, 2004.

http://en.wikipedia.org/wiki/Climate#Climate_change

http://en.wikipedia.org/wiki/Weather#Weather_modification_and_human_impact

http://nsidc.org/arcticmet/basics/weather_vs_climate.html

http://shodhganga.inflibnet.ac.in/bitstream/10603/69214/11/11_chapter%201.pdf http://www.alevelgeography.com/water-balance/

http://www.mpimet.mpg.de/en/presse/faq-s/was-ist-der-unterschied-zwischen-wetter-und-klima.html

http://www.nasa.gov/mission_pages/noaa-n/climate/climate_weather.html

Koppen, W. 1936. Das geographische System der klimate. In Handbuch der Klimatologie, Vol. 1, Part C., ed. W. Kbppen and R. Geiger. Berlin : Gebruder Borntrager

Kottek, Markus; Grieser, Jürgen; Beck, Christoph; Rudolf, Bruno; Rubel, Franz (2006). "World Map of the Köppen-Geiger climate classification updated" (PDF). Meteorologische Zeitschrift. 15 (3):259–263. Bibcode:2006MetZe..15..259K. doi:10.1127/0941-2948/2006/0130

Luke Howard, English chemist and meteoroligist, early 19th century ; Retrieved 23 September 2008

Lutgens, F. K. et al. The Atmosphere: An Introduction to Meterorology, 8th ed. Englewood Cliffs, NJ: Prentice Hall, 2001.

Mukherjee, SP; Climatic Classification of Koppen; University Deptt. Ranchi : Advanced Climatology, P. G. Semester-II Paper Code (GEO 201)

Papadakis, J. 1975: Climates of world and their agricultural potentialities. Buenos Aires

Ruddiman, W. F. Earth's Climate: Past and Future, 2nd ed. New York, NY: W. H. Freeman and Company, 2008.

RusselL, J.S. 1978: Classification of Climate and the potential usefulness of pattern analysis techniques in agro-climatological research. Presented at the International Agro-climatology Workshop, 22-24 November 1978, ICRISAT, Hyderabad, India

Stevens, A. N.P. (2011) Factors Affecting Global Climate. Nature Education Knowledge 3(10):18

Thornthwaite, C.W. 1948: An approach towards a rational classification of climate. Geographical Review 38:55-94

Thornthwaite, C.W., and Mather, J.R. 1955. The water balance. Publications in climatology, Vol.8(No.1) , Drexel Institute of Technology, Laboratory of Climatology, Centerton, N.J. 104 pp.

Trewarta, G T. (1968), An Introduction to Climate, International Student Edition, Mc Graw Hill Kogakusha Ltd, Tokyo. Oliver, J.E.

2

Climate of India & Odisha

Modern agriculture requires certain precise information on pattern of rainfall, temperature, occurrence of drought and flood etc. for formulating required strategy for crop production. It also requires knowledge about agro-climatic conditions of the region for deciding a planned cropping pattern. Therefore climatic classification and delineation of agro-climatic zones would be imperative for a country or state.

2.1 Climatic Regions of India

In India the tropic of Cancer passes through the central part in the East-West direction. Hence the Northern part of India lies in the sub-tropical and temperate zone and the part lying in the South of the Tropic of Cancer remains in the tropical zone. Hence two major climatic regions of India are temperate or continental India and tropical India. The temperate region is sub-divided into five regions, namely Himalayan region (heavy rainfall), North-western region (moderate rainfall), Arid low land, Region of moderate rainfall and the transitional zone. The tropical zone has also been sub-divided into six regions, namely Region of very heavy rainfall, Region of heavy rainfall, Region of moderate rainfall, The Konkan Coast, The Malabar Coast, and Tamil Nadu.

2.1.1 Climatic Regions of Temperate or Continental India

The Himalayan Region: This region covers Jammu and Kashmir, Himachal Pradesh, large parts of Uttaranchal, the north hill region of West Bengal, Sikkim and Arunachal Pradesh. The winter and summer temperatures of the region are 4°-7°C and 13-18°C , respectively. Annual rainfall is more in eastern states with average of 200 cm or more.

The North-western Region: It covers the northern parts of Punjab and southern parts of Jammu and Kashmir. The winter and summer temperature of the region are 16°C and 24°C, respectively. Annual average rainfall is below 200cm.

The Arid Lowland: The region covers vast area of Thar Desert of Rajasthan, south western part of Haryana and Kuchchh of Gujarat. The average temperature in winter varies from 16° to 24°C which goes up to 48°C in summer. There is low rainfall below 40 cm.

The region of moderate rainfall: Parts of Punjab, Haryana, western Uttar Pradesh, Union Territory of Delhi, north-west Plateau area of Madhya Pradesh and eastern Rajasthan are covered by this region. The annual rainfall of the region is 40 to 80 cm. Temperature in winter varies between 15°-19°C and it remains within 30°-35°C during summer.

The Transitional Zone: Eastern Uttar Pradesh and Bihar comes under the transitional zone. The temperatures in January and July vary between 15°-19°C and 30°-35°C. The rainfall remains within 100-150 cm in the year.

2.1.2 Climatic Regions of Tropical India

Region of very heavy rainfall: With more than 200 cm of annual rainfall the region includes large parts of Meghalaya, Assam, Tripura, Mizoram and Nagaland. Temperature remains around 18°C in January and rises to 32°- 35°C in July. Cherrapunji and Mawsynram in Meghalaya receive the highest annual rainfall of 1102 cm and 1221 cm, respectively. Other places get more than 200 cm of rainfall.

Region of heavy rainfall: Chhattisgarh, Jharkhand, Gangetic West Bengal, Odisha and north-eastern Andhra Pradesh come under this region which receives 100-200 cm annual rainfall. The winter and summer temperatures range from 18°-24°C and 29°-35°C, respectively.

Region of moderate rainfall: The region lies between Western and Eastern Ghats which receives annual rainfall of 50-100 cm. The average temperature in winter is 18°-24°C which rises to 32°C in summer.

The Konkan Coast: This region extends from Mumbai in the north to Goa. The annual rainfall is over 200 cm . The temperature varies from 24°-27°C.

The Malabar Coast: The region extends from Goa to Kanniyakumari and receives heavy annual rainfall of over 250 cm. The rainfall is mainly brought by the south-west monsoon winds and continues for about nine months in a year. The temperature remains in the vicinity of 27°C .

Tamil Nadu: The region covers Tamil Nadu and adjoining areas of Andhra Pradesh. The rainfall varies from 100 to 150 cm and is mainly caused by the retreating monsoons from north-east during November and December. The temperature is around 24°C without much change in summer and winter temperature.

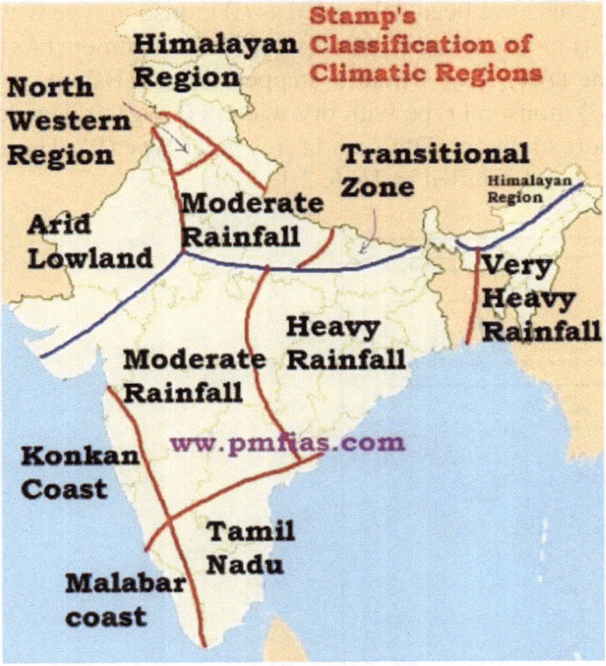

Fig. 2.1: Climatic Regions of India (After Stamp)

2.2 Classification of Climate in India

Classification of climate in India has been made by using temperature, rainfall, evapotranspiration etc. Wladimir Peter Koppen classified the climate in 1884 by using temperature and precipitation averages as these are two important factors for spread of vegetation. Thornthwaite (1948) used average monthly potential evapotranspiration and corresponding rainfall for classification of climate. Later Thornthwaite and Mather (1955) improved the classification with "Rational classification" which introduced various degrees of deficit and surplus of water. Papadakis (1966, 1975) used water balance techniques to classify climate. Chowdhary and Sarwade (1980) suggested combination of humid and arid months for a satisfactory classification of climate. Hargreaves (1971) used monthly moisture availability index (MAI) which he defined as the ratio of monthly precipitation at 75% probability level to monthly potential evapotranspiration.

2.2.1 Koppen's Scheme

Koppen's classification of climate regions of India is an empirical classification based on temperature and precipitation data. On the basis of rainfall and temperature the climate regions have been delineated as (i) monsoon type with short dry season (Amw), (ii) monsoon type with dry season in summers (As), (iii) tropical savannah type (Aw), (iv) semiarid steppe climate (Bshw), (v) hot desert type (Bwhw) (vi) monsoon type with dry winters (Cwg), (vii) cold humid winter type with short summers (Dfc) , and (viii) polar type (E). These regions are shown in Fig.2.2 and detailed in Table 2.1.

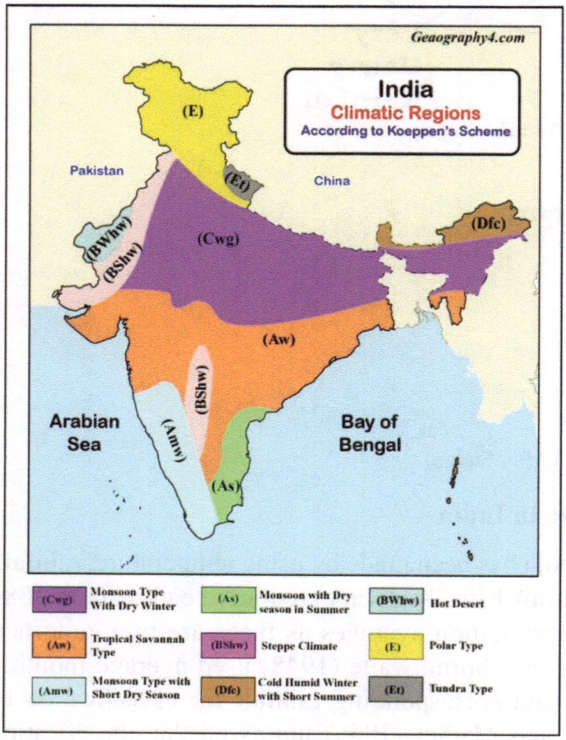

Fig. 2.2: Climate regions of India by Koeppen

According to Koppen's scheme the Amw type of climate prevails over the western coast of India and south of Goa. Monsoon type of dry climate (As) is found along the Coromandel Coast. The dry climate (Bshw) prevails over interior peninsula, Rajasthan and parts of Haryana. The extreme western Rajasthan experiences hot type of climate (Bwhw). Tropical savannah type of climate (Aw) is seen in most of the peninsular plateau. Warm temperate type of climate with dry winters (Cwg) is found in plains of India. The north-eastern India experiences Dfc type of climate with cold and humid winter and short summer. Tundra type of climate (Et) is found in mountains of Uttarakhand. Kashmir and the adjoining mountain ranges have a polar type of climate (E).

Table 2.1: Climate Regions of India (Koeppen's scheme)

Climate type	Region	Annual rainfall
Amw (Monsoon type with short dry winter season)	Western coast and south of Goa.	Over 300 cm
As (Monsoon type with dry season in high sun period)	Coromandel coast - Coastal Tamil Nadu and bordering AP and Telengana regions	75 – 100 cm (wet winters, dry summers)
Aw (Tropical Savanna type)	Most of the peninsular plateau except Coromandel and Malabar coastal regions.	75cm
BShw (Semi-arid)	Interior peninsula, Rajasthan and parts of Haryana	12cm to 25cm
BWhw (Hot desert type)	Extreme western Rajasthan	Less than 12cm
Cwg (Warm temperate type with dry winters)	Plains of India	100cm to 200cm
Dfc (Cold, Humid winters type with shorter summer)	North-eastern India	Approximately 200cm
Et (Tundra Type)	Uttarakhand's mountainous terrain.	The amount of rain that falls fluctuates from year to year.
E (Polar Type)	Higher portions of Jammu and Kashmir and Himachal Pradesh	Snow is the most common kind of precipitation.

2.2.2 Thornthwaite's Scheme

Thornthwaite's climatic classification scheme in India is based on water balance. The climate is classified according to water deficit and water surplus expressing moist and dry climates. According to Thornthwaite's method, the climatic regions of India are identified as (i) per-humid (A), (ii) humid (B), (iii) moist sub-humid (C2), (iv) dry sub-humid (Cj), (v) semi-arid (D), and (vi) arid (E). The per-humid region (A) lies along the west coast of India, south of Goa and some parts of north-eastern India has a moist climate. The humid climate (B) prevails along west coast, northern part of West Bengal, and the neighbouring parts of north-eastern India. Moist sub-humid (C2) climate prevails along the Western Ghats and also over Odisha and West Bengal. The Ganga valley and north-eastern parts of Central India have a dry sub-humid type of climate (C1). Semi-arid type of climate (D) is found in the peninsular interior, western Madhya Pradesh, Haryana and Punjab. Saurashtra, Kuchchh and Rajasthan experience an arid climate (E). The climate classification of India according to Thornthwaite's scheme is shown in Fig, 2.3.

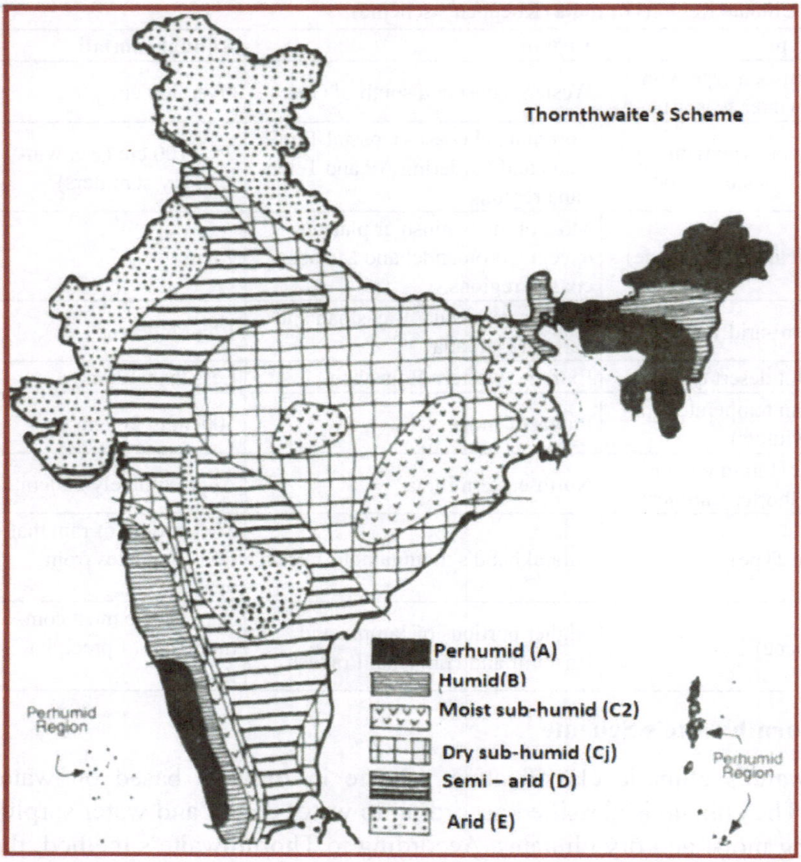

Fig. 2.3: Climatic regions of India (Thornthwaite's Scheme)

2.2.3 Trewartha's Scheme

Climatic classification by Trewarth's scheme is a modification of Koeppen's scheme. Based on vegetation, agriculture and geographical regions Trewartha's classification has four major climatic groups A, B, C, and H that are further subdivided into seven climatic types as shown in Table 2.2 and Fig. 2.4.

Table 2.2: Climatic Classification of India as per Trewartha's Scheme

Type	Climate group	Climate type
A	Tropical rainy climate Group	1. Am – Tropical Monsoon- This type of Climate covers the Indian region of Western Ghat, Western Nagaland, and Tripura. 2. Aw- Tropical savannah – The region covers the whole of the Deccan Plateau except a narrow strip of rain – shadow area in the East of Western Ghats, north-eastern Gujarat, Southern Madhya Pradesh, Southern Bihar, Odisha, Andhra Pradesh, and Tamil Nadu.
B	Dry climatic group	1. BS – Tropical Steppe (semi-arid)- The region covers interior Karnataka, Central Maharashtra, Western Andhra Pradesh, and interior Tamil Nadu. 2. Bsh- Subtropical Steppe – It covers an area from Punjab to Kachchh 3. Bwh -Subtropical Desert – Western Rajasthan and Kachchh
C	Humid Mesothermal climatic group	1. Caw – Subtropical Humid (Dry winters)- Punjab foothills, Uttar Pradesh, Bihar, West Bengal, Assam, and Arunachala Pradesh
H	Mountain Climate	1 Mountain area of Kashmir and Arunachal Pradesh

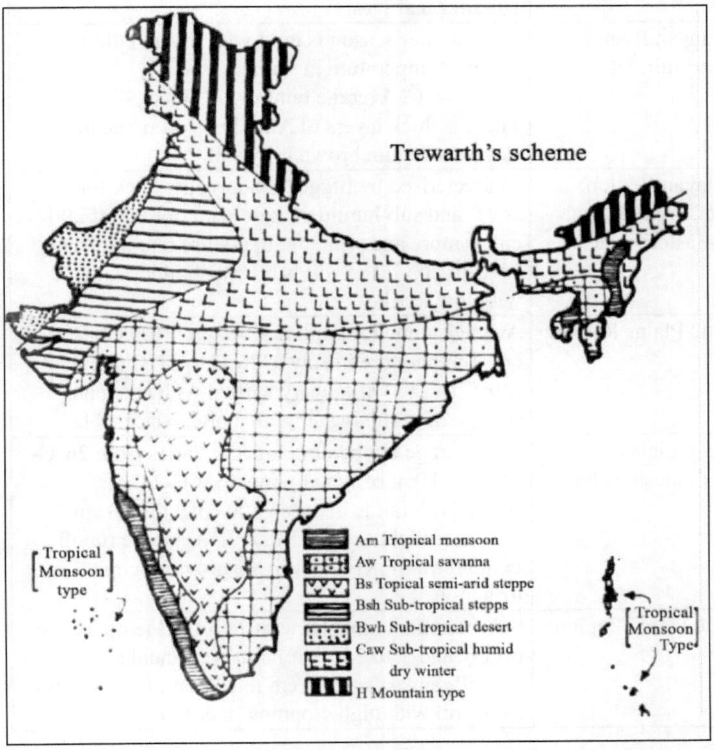

Fig. 2.4: Climatic classification of India as per Trewartha's scheme

2.3 Agro-climatic Zones of India

For the purpose of effective crop planning, agro-climatic zones of India have been delineated based on climate and suitability of different crops by Indian Council of Agricultural Research (ICAR), Planning Commission, National Agricultural Research Project (NARP) under ICAR, National Bureau of Soil Survey and Land Use Planning (Agro -ecological regions by the NBSS & LUP).

(i) Agro-climatic regions by the erstwhile Planning Commission

The Planning Commission of India has divided the country into fifteen broad agro-climatic zones based on physiography, soils, geological formation, climate, cropping patterns, and development of irrigation and mineral resources for broad agricultural planning and developing future strategies in seventh plan period. The main objective was to integrate plans of the agro-climatic regions with the state and national plans to enable policy development based on techno-agro-climatic considerations.

Table 2.3: Agro-climatic zones in India by Planning Commission

Zone	Regions	Distinct features
1	Western Himalayan Region: Jammu and Kashmir, Uttar Pradesh	The summer season is mild with severe cold in winter. Temperature in January is 0°C to -4°C. Average rainfall is 150 cm. There is thick layers of Alluvium soil while hill slopes have thin brown hilly soil.
2	Eastern Himalayan Region: Assam, Sikkim, West Bengal and all North-Eastern states	Characterised by rugged topography, thick forest cover and sub-humid climate with rainfall of 200 cm or more and winter temperature ranging from 11°C to 20°C. The soil is brownish, thick layered with low fertility.
3	Lower Gangetic Plains Region: West Bengal	Average annual rainfall lies between 100 cm-200 cm. Temperature in July is 26°C to 41°C and it is 9°C to 24°C in January. Gifted with adequate storage of groundwater with a high water table.
4	Middle Gangetic Plains Region: Uttar Pradesh, Bihar	The average temperature of July varies from 26°C-41°C and that of January month 9°C-24°C. Annual rainfall is between 100 cm and 200 cm. Vast potential of groundwater and surface runoff in the form of perennial rivers provide scope for irrigation.
5	Upper Gangetic Plains Region: Uttar Pradesh	Sub-humid continental with 26°-41°C temperature in July and 7°to 23°C in January. Annual average rainfall ranges from 75 cm to 150 cm. Ideal sandy-loam soil with high cropping intensity.

Zone	Regions	Distinct features
6	Trans-Gangetic Plains Region: Punjab, Haryana, Delhi and Rajasthan	Semi-arid climate with temperature of 26°C to 42°C in July and 7°C to 22°C in January. The average rainfall is 70 cm to 125 cm.
7	Eastern Plateau and Hills Region: Maharashtra, Uttar Pradesh, Odisha and West Bengal	The temperature of July is 26°C to 34°C and 10°C to 27° C in January. Annual rainfall ranges from 80 cm to 150 cm. Red and yellow soils with occasional patches of laterites alluvial are seen.
8	Central Plateau and Hills Region: MP, Rajasthan, Uttar Pradesh	Semi-arid climate in the western part and sub-humid in the eastern part with temperature of 26°C to 40°C in July and 7°C to 24°C in January are seen in the region. Average annual rainfall from 50 cm to 100 cm with mixed red, yellow and black soils are found.
9	Western Plateau and Hills Region: Maharashtra, Madhya Pradesh and Rajasthan	Temperature of July remains between 24°C and 41 °C while temperature of January is 6°C to 23°C . Annual d average rainfall is 25 cm to 75 cm.
10	Southern Plateau and Hills Region: Andhra Pradesh, Karnataka, Tamil Nadu	Semi-arid climate with temperature of July ranges from 26°C to 42°C and that of January ranges from 13°C to 21°C. Annual rainfall of 50 cm to 100 cm with only 50 per cent of the area cultivated.
11	East Coast Plains and Hills Region: Odisha, Andhra Pradesh, Tamil Nadu and Pondicherry	Sub-humid maritime climate with temperature of 26°C to 32°C in May and 20°C to 29°C in January. Annual rainfall is 75 cm to150 cm. Alluvial, loam and clay soils with alkalinity are mostly seen.
12	West Coast Plains and Ghat Region: Tamil Nadu, Kerala, Goa, Karnaṭaka, Maharashtra	Humid region with annual rainfall above 200 cm and average temperatures of 26°C-32°C in July and 19°C-28°C in January.
13	Gujarat Plains and Hills Region: Gujarat	Arid and semi-arid climate with average annual rainfall 50 cm to 100 cm. Temperature of July remains between 26°C to 42°C and 13°C to 29°C in January. Soils are alluvium in the coastal plains, and red and yellow soils in Jamnagar.
14	Western Dry Region: Rajasthan	Hot sandy desert climate with erratic rainfall below 25 cm. There is high evaporation with temperature of 28°C to 45°C in June and 5°C to 22°C in January. There are no such perennial rivers, and vegetation is scanty.
15	The Islands Region: Andaman and Nicobar, Lakshadweep	Annual rainfall is less than 300 cm and the temperatures of Port Blair being 30°C and 25°C respectively. The dominant soils vary from sandy along the coast to clayey loam in valleys and lower slopes.

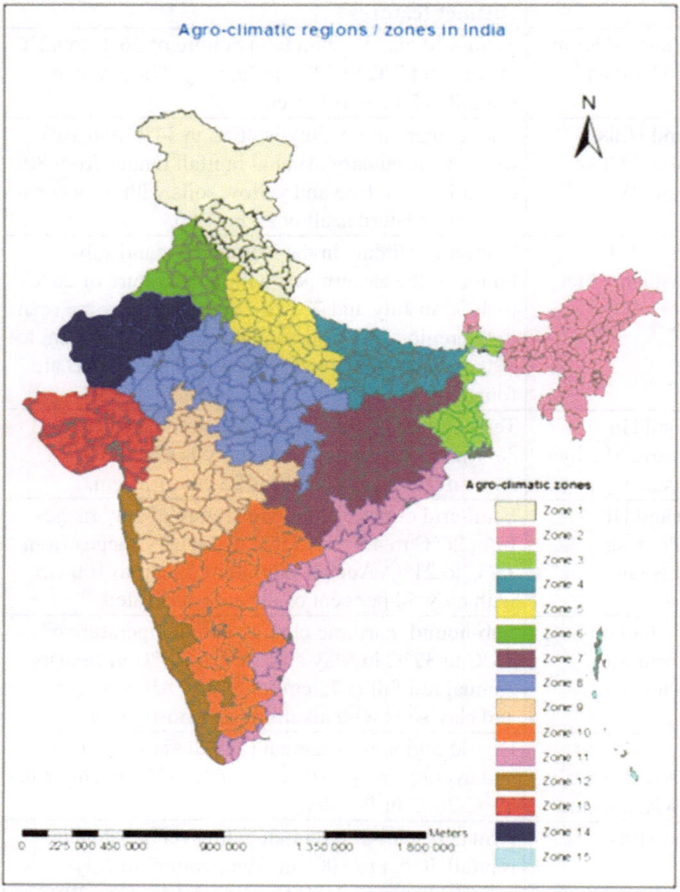

Fig. 2.5: Agro-climatic zones of India (Planning Commission)

(ii) Agro-climatic zones of India identified for National Agricultural Research project (NARP) of ICAR

Indian Council of Agricultural Research (ICAR) identified 129 agro-climatic zones to establish zonal research stations for conducting research in specific agro-ecological situations. The thrust was given for analysing agro-ecological conditions and appropriate cropping patterns basing on the major problems, natural resources, major crops, farming systems and production constraints. Table 2.4 gives the number of sub-zones in each agro-climatic zone.

Table 2.4: Agro-climatic sub-zones by ICAR

S.No	Name of ACZ	Number of sub-zones
1	Andhra Pradesh	7
2	Assam	6
3	Bihar	6
4	Gujarat	8
5	Haryana	2
6	Himachal Pradesh	4
7	Jammu & Kashmir	4
8	Karnataka	10
9	Kerala	8
10	Madhya Pradesh	12
11	Maharashtra	9
12	NE Hill Regions	6
13	Odisha	10
14	Punjab	5
15	Rajsthan	9
16	Tamil Nadu	7
17	Uttar Pradesh	10
18	West Bengal	6
	Total	129

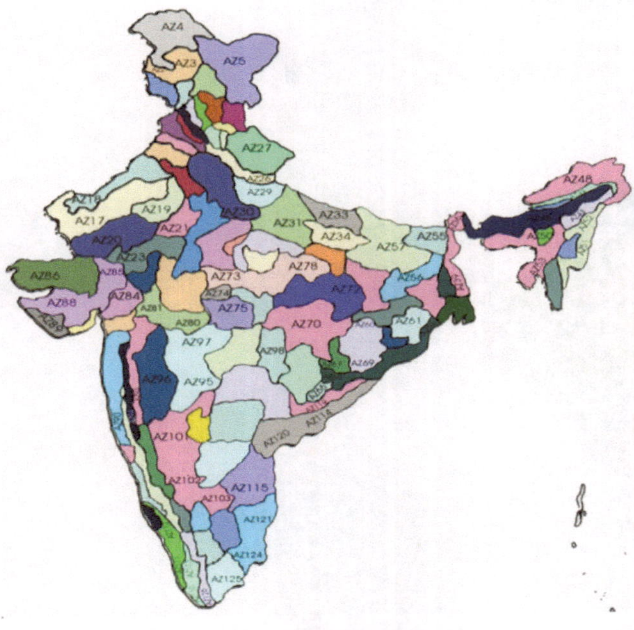

Fig. 2.6: Agro-climatic zones of India (NARP)

(iii)Agro-ecological regions (National Bureau of Soil Survey & Land Use Planning (NBSS & LUP)

The National Bureau of Soil Survey and Land Use Planning (NBSS&LUP; ICAR) adopted the length of growing season (LGS) for identifying the agro-ecological zones/regions of India. LGS considers the soil water balance as a function of moisture availability in a landform. By superimposing bio-climate and LGS on soil-scape, 20 Agro-ecological regions were delineated. Considering the crop duration classes such as short (< 90 days), medium (90-150 days), long (150-180 days), relay cropping (180-210 days) and double cropping (>210 days) five LGS classes were developed. In consideration of soil depth and available water capacity, these 20 zones were further sub-divided in to 60 agro-ecological sub-regions (AESR).

The agro-ecological zones/regions developed by NBSS&LUP) are: (1) Western Himalayas, (2)Western Plain, (3)Deccan Plateau, (4)Northern Plain, (5)Central Malwa Highlands,(6) Gujarat Plains, (7) Deccan Plateau, (8) Deccan (Telengana) Plateau, (9)Eastern Ghats,(9) Northern Plain, (10)Central Highlands, (11)Eastern Plateau (Chattisgarh), (12) Eastern (Chotanagpur) Plateau and Eastern Ghats, (13)Eastern Plain, (14)Western Himalayas, (15) Bengal and Assam plains, (16)Eastern Himalayas, (17)North Eastern Hills, (18) Eastern Coastal Plain, (19) Western Ghats and Coastal Plain, and (20). Island of Andaman Nicobar and Lakshadweep.

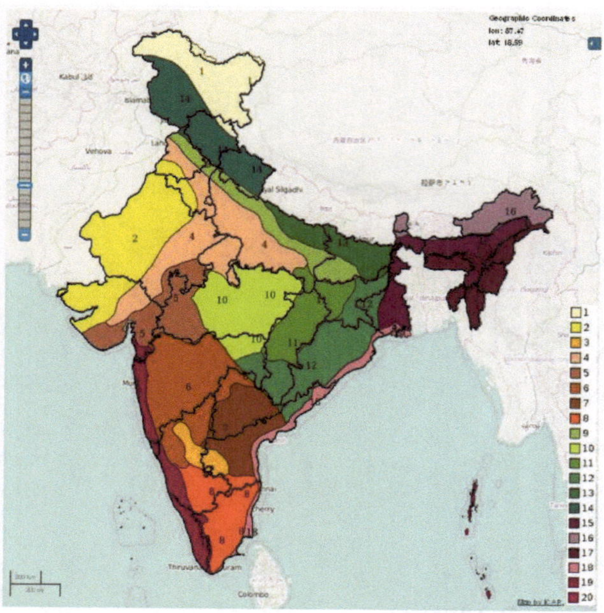

Fig. 2.7: Agro-ecological zones of India (NBSS& LUP)

2.4 Climate of Odisha

The state lies in the north-eastern part of the Indian Peninsula, approximately between latitudes 22^0 36'N and 17^0 49'N and longitudes 81^0 36'E and 87^0 18'E. and 870 18'E. Odisha has a geographical area of 15.58 million ha with a coastal line of 480 km and 30 districts. The state may be broadly divided into four geographical regions, viz. the northern plateau, central river basins, eastern hills and coastal plains.

Geographical region	**Districts**
Northern plateau	Mayurbhanj, Keonjhar and Sundargarh
Central river basin	Bolangir, Sonepur, Sambalpur, Deogarh, Bargarh, Jharsuguda,Dhenkanal , Angul , and a part of Cuttack district
Eastern hills	Koraput, Rayagada, Nawarangpur, Malkangiri, Kalahandi, Nawapara, Gajapati and a part of Ganjam district
Coastal plains	Balasore, Bhadrak, Kendrapara, Jagatsinghpur, Jajpur, Puri, Khurda, Nayagarh and a portion of Ganjam and Cuttack districts.

2.4.1 Koppen's classification

Koppen's classification is based on the variation of temperature and rainfall. The state as a whole mainly comes under the climatic type: Tropical Savanna Hot; seasonally dry (usually winter) (Aw) except parts of Raygada, Kandhamal, Boudh, Kalahandi,Malkanagiri and Koraput. The climatic classification is as follows:

(i) Tropical Savanna (hot, seasonally dry in winter: (w): All districts except a part of Raygada, Koraput and Kandhamal, Boudh and Malkanagiri

(ii) Sub-tropical monsoon, mild and dry winter with hot summer (Aw and Cwa): Raygada and part of Kalahandi, Koraput and Nabarangpur

(iii) Sub-tropical monsoon, mild winter and hot summer (Cwa); Kandhamal, part of Koraput and Malkanagiri

2.4.2 Thornthwaite (1948) and Thornthwaite & Mather (1955)

Thornthwaite introduced the concept of potential evapotranspiration (PET) and compared it with precipitation to calculate moisture index (Im).He suggested to calculate moisture deficit (d) and moisture surplus (s) period. The PET was calculated from mean monthly temperature and annual temperature index. Subsequently Thornthwaite and Mather (1955) revised the method of calculating aridity and humidity index and revised the moisture index (Im).

With the revised moisture index $\left(\text{Im} = \frac{P - PET}{PET}\right)$, the following climate types were suggested Table 2.5.

Table. 2.5: Moisture deficit index and climate type corresponding to Im.

Climate type	**Moisture Index**
A-Per-humid	>100
B_4-Humid	80-100
B_3-Humid	60-80
B_2-Humid	40-60
B_1-Humid	20-40
C_2-Moist sub-humid	0-20
C_1- Dry sub-humid	-33.3-0
D-Semi arid	-66.7-33.3
E -Arid	< -66.7

Source: Lenka, D: Agriculture in Odisha, 2001

Basing on the data on precipitation, mean annual temperature and moisture index of all the undivided districts of Odisha, Koraput, Keonjhar and Kandhamal come under moist sub-humid while all other 10 undivided districts come under dry sub-humid type of climate (Lenka, D, 2001)

Aridity Index & Precipitation Effective Index

The aridity index is calculated by the formula R x N/E, where R= Annual rainfall in cm, N=number of rainy days and E=Annual evaporation in cm.

Precipitation Effective Index (PEI) is calculated as:

$$PEI = \sum_{n=1}^{12} 115\left(\frac{P}{T} - 10\right)^{1.111}$$

Where P=annual rainfall (inches), T=Annual temperature (^{0}F) and N=Number of months.

The aridity index and PEI have been given in Table 2.7.

Table 2.6: AI and PEI of undivided districts of Odisha

District	**Aridity Index**	**Precipitation Effective Index**
Mayurbhanj	76(80)	99
Keonjhar	66(72)	99
Sundargarh	73(84)	98
Dhenkanal	65(66)	86
Sambalpur	64(69)	91
Bolangir	57(63)	88
Koraput	72(73)	105
Phulbani	65(73)	99
Kalahandi	58(58)	83

District	Aridity Index	Precipitation Effective Index
Balasore	70(73)	96
Cuttack	67(69)	89
Puri	67(65)	86
Ganjam	53(65)	86
Mean	61(69)	92

(NB: Figures in parentheses indicate the value of rainfall from 1901-1950)

In consideration of the aridity index calculated taking the rainfall from 1901-50 into consideration, the districts of Odisha have been grouped under following types:

Aridity Index	Undivided districts
Low (50-60)	Kalahandi, Bolangir , Ganjam
Medium (60-70)	Keonjhar, Dhenkanal, Sambalpur, Phulbani, Balasore, Cuttack, Puri
High (>70)	Mayurbhanj, Koraput, Sundargarh

The classification according to temperature is indicated below.

Class	Type	Temperature
A	Very hot	>28^0 C
B	Hot	25-28^0C
C	Mild	20-25^0C
D	Cold	10-20^0C
E	Very cold	<10^0C

According to moisture index and temperature classes, the undivided districts of Odisha have been classified into four broad types (Table 2.8). The state ,in general, is slightly dry and mildly hot.

Table 2.7: Classification of undivided districts of Odisha into Temperature and Moisture classes

Type	Mild	Mild Hot	Hot
Slightly dry		Sundargarh	Mayurbhanj, Sambalpur, Cuttack, Puri, Dhenkanal, Ganjam
Dry			Bolangir
Slightly moist	Phulbani		Kalahandi, Balasore
Moist		Keonjhar, Koraput	

Source: Lenka, D:Agriculture in Orissa, 2001

2.5 Safe Growing Period

The beginning of the growing period starts when the precipitation is equal to 0.5 PET. This period meets the demand of crop and replenishes the moisture deficit of the soil profile. The period when P is less than 0.5 PET, it is termed as the end of rainy season but not the end of the growing season since some moisture is stored in soil. The growing period ends when the precipitation decreases further to less than 0.5 PET plus the time taken to use the stored soil moisture (usually 100 mm). The difference between the field capacity (FC= -0.33 bar) and permanent wilting point (PWP= -15 bar) is regarded as the relatively available moisture. Basing on this concept ten growing period classes have been proposed.

Class	Growing period in days
1	Less than 60
2	60-90
3	90-120
4	120-150
5	150-180
6	180-210
7	210-240
8	240-270
9	270-300
10	>300

Source: Lenka, D:Agriculture in Orissa, 2001

2.6 Agro-climatic Zones of Odisha

There are ten agro-climatic zones in Odisha which are not necessarily co-terminus with the district boundaries. The coverage of the ACZs is indicated in Table 2.9.

Table 2.8: Particulars of ACZs of Odisha

S.No	ACZ	Agricultural Districts	Climate	Mean Rainfall	Mean Max Temp	Mean Min Temp	Broad soil group
1	North-Western Plateau	Sundargarh, parts of Deogarh, Sambalpur & Jharsuguda	Hot & moist sub-humid	1600	38.0	15.0	Red, Brown forest, Red & Yellow, Mixed Red & Black
2	North-Central Plateau	Mayurbhanj, major parts of Keonjhar, (except Anandapur & Ghasipura)	Hot & moist sub-humid	1534	36.6	11.1	Lateritic, Red & Yellow, Mixed Red & Black
3	North-Eastern Coastal Plain	Balasore, Bhadrak, parts of Jajpur & Hatdihi block of Keonjhar	Moist Sub-humid	1568	36.0	14.8	Red, Lateritic, Deltaic alluvial, Coastal alluvial & Saline
4	East & South Eastern Coastal Plain	Kendrapara, Khurda, Jagatsinghpur, part of Cuttack, Puri, Nayagarh & part of Ganjam	Hot and humid	1577	39.0	11.5	Saline, Lateritic, Alluvial, Red & Mixed red & Black
5	North-Eastern Ghat	Kandhamal, Rayagada, Aska of Ganjam and Paralakhemundi of Gajapati	Hot and Moist, sub-humid	1597	37.0	10.4	Brown forest, Lateritic Alluvial, Red, Mixed Red & Black
6	Eastern Ghat High Land	Major parts of Koraput, Nabarangpur	Warm and humid	1522	34.1	7.5	Red, Mixed Red & Black, Mixed Red & Yellow
7	South-Eastern Ghat	Malkangiri & part of Koraput	Warm and Humid	1710	34.1	13.2	Red, Lateritic, Black
8	Western Undulating	Kalahandi & Nuapada	Hot and moist sub-humid	1352	37.8	11.9	Red, Mixed Red & Black , Black
9	West- Central Table Land	Bargarh, Bolangir, Boudh, Sonepur, parts of Sambalpur & Jharsuguda	Hot & moist sub-humid	1614	40.0	12.4	Red & Yellow, Red & Black, Black, Brown forest, Lateritic
10	Mid-Central Table Land	Angul, Dhenkanal, Athagarh and Sukinda block of Jajpur	Hot & moist sub-humid	1421	38.7	14.0	Alluvial, Red, Laterite Mixed Red & Black

Source: Odisha Agriculture Statistics, DA&FP, Odisha

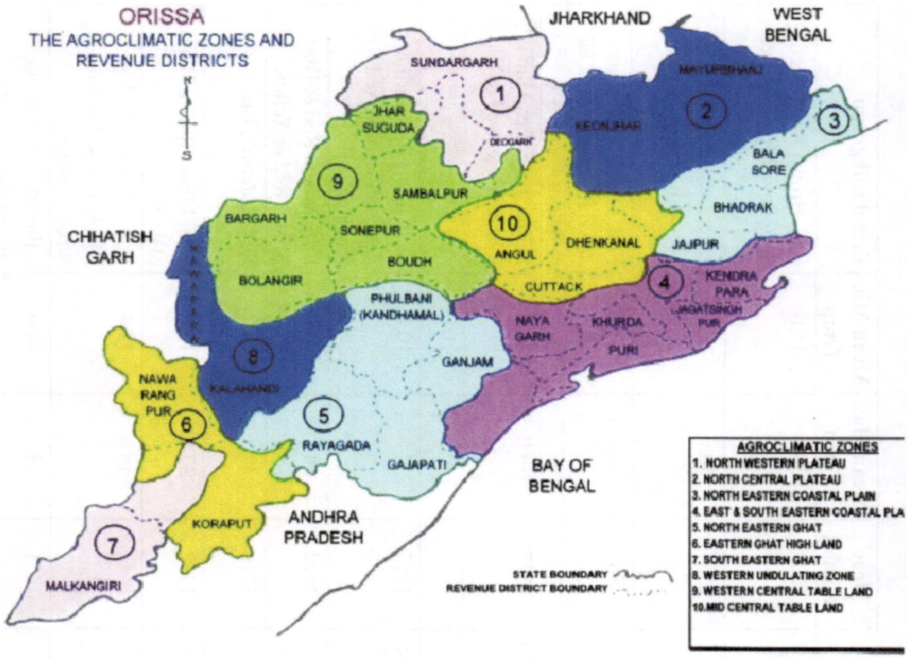

Fig. 2.8: Agro-climatic Zones of Odisha

References

https://docplayer.net/104047944-Sixth-semester-geography-notes-unit-i-by-dr-rafiq-ahmad-hajam-deptt-of-geography-gdc-boys-anantnag-cell-no.html

https://www.pmfias.com/climatic-regions-of-india-stamps-koeppens-classification

Koeppen, W 1936. Das geographische System der klimate. ln Handbuch der KlimetologieK , V o l . 1, Part C., ed. W. Koeppen and R. Geiger. B e r l i n : Gebruder Borntrager

Lenka, D (2001); Agriculture in Orissa; Kalyani Publishers

Odisha Agriculture Staitics, Director of Agriculture & Food production, Odisha

Papadakis,J. 1966. Crop ecology survey in West Africa . Buenos Aires

Papadakis, J. 1975. Climates of the world and their agricultural potentialitiues. Buenos Aires.

Penman, H.L. 1948. Natural evaporation from open water, bare s o i l and grass. Proceedings, Royal Society, Series A, 193:120-145.

Sinha, D (1989): Agro-climatic zones of India categorised by the Planning Commission

Thornthwaite, C.W. and Mather, J. R., The water balance. In Climatology, Drexel Institute of Technology, Laboratory of Climatology, Centerton, NJ, 1955, vol. 8, No. 1, p. 104.

Trewarta, G T. (1968), An Introduction to Climate, International Student Edition, Mc Graw Hill Kogakusha Ltd, Tokyo.

Virmani, S.M et al (ICRISAT): Climatic Classification of the Semi-arid tropics in relation to farming System Research.

3

Climate and Crop Production

3.1 Climatic Factors and Crop Growth

Crop growth depends on internal and external factors. Internal factors are mostly genetic or hereditary factors which include high yielding ability, resistance to lodging, tolerance to adverse conditions such as drought, flood and salinity, tolerance to insect pests and diseases, and grain quality. The external factors include climate, edaphic, biotic, physiographic and socio-economic factors.

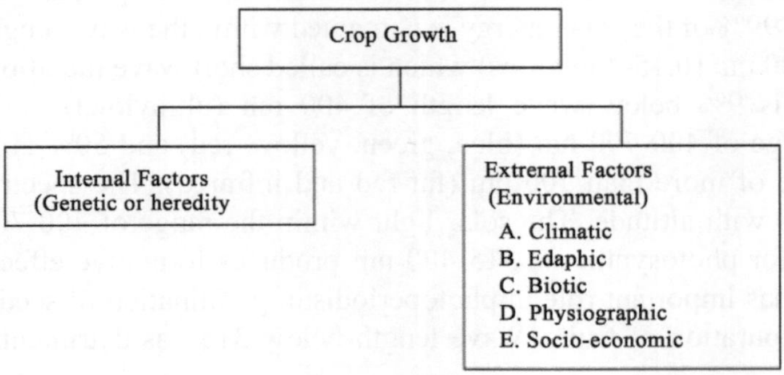

Climate influences the yield potential of a crop. Nearly 50% of crop yield is attributed to the impact of climatic factors. The climatic factors are mostly abiotic that affect plant growth and development. Such factors that affect plant growth, development and crop productivity are water, temperature, solar radiation, atmospheric humidity, wind velocity and atmospheric gases. However, solar radiation, temperature and rainfall are most important factors, which influence the growth, development and yield of crops.

3.1.1 Solar Radiation

Solar radiation is the source of all heat energy (about 99.97% of total energy of the Earth) which is required for all physical processes taking place in Earth-atmosphere system. Solar radiation provides light which regulates the seed germination, leaf expansion, growth of shoot, flowering, fruiting and other physiological functions in plants. Effect of solar radiation on plants can be divided into four important categories.

(i) **Thermal effect:** Bulk of solar radiation absorbed by plants is converted to heat energy which is used for transpiration. Temperature obtained from solar radiation is dealt under temperature (3.1.3).

(ii) **Photosynthetic effect:** About 28% of solar energy is utilised for photosynthesis in plants for production of carbohydrates. Light quality, intensity and day length are important factors regulating growth and development of plants.

(iii) **Photoperiodic effect:** Solar radiation regulates and controls the flowering and vernalisation in plants according to length of light period.

(iv) **Other effects:** Solar radiation influences assimilation of plant nutrients and dry matter distribution.

Solar energy is an electromagnetic phenomenon which bears the characteristics to a body at 6000^{0}K and is the result of thermonuclear fusion of many hydrogen atoms. More than 99% of the solar energy is contained within the wave length region of 150-4000 nm (0.15-4 microns) which is called short wave radiation. The distribution is 9% below wave length of 400 nm (ultraviolet), 45% within visible range of 400-700 nm (blue, green, yellow, red) and 50% is in the infrared range of more than 700 nm (far-red and infrared). The spectral distribution varies with altitude. The solar light within the range of 400-700 nm are effective for photosynthesis, 315-400 nm produces formative effects and 700-920 nm has important role in photoperiodism, germination of seeds, flowering and colouration of fruits. Wave length below 315 has detrimental effect on plants.

0-400 nm	400-700 nm			700-920 nm	920-1900nm
Ultraviolet	Blue	Green/Yellow	Red	Far-red	Infrared
	400-510	510-610	610-700		

Wavelength (nm)10-380 450 495 570 590 620 750-1900
Colour Violet Green Orange Infrared
Ultraviolet Blue Yellow Red

Fig.3.1: Wave length of light (nm)

The solar radiation on the top of the atmosphere depends on latitude, season and day time hour. Solar ration falling on the Earth is also reflected which is called albedo. Albedo is measured as the ratio of incident solar radiation reflected by a body and expressed in percentage. The emissivity varies with the nature of the surface. The values of E given by Geiger (1961) and Brooks(1959) are indicated in Table.3.1.

Table 3.1: Value of E and Albedo

Surface	Infrared emissivity	Albedo (less than 4 micron)
Water surface	92-96	6-9
Snow	82-99	75-95
Soil dry, dark	90	5-15
Wet grass	-	33-37
Dry grass	-	15-28
Earth	-	34-42

During photosynthesis carbon dioxide is reduced to carbohydrates in presence of catalytic enzymes such as phosphoenol pyruvate carboxylase and ribuse bisphosphate carboxylase which are regulated by light intensity. Different pigments required for photosynthesis are also produced in presence of light. Since solar radiation influences crop growth directly, its reduction will appreciably reduce agricultural productivity. The accompanied increase in minimum temperatures increases respiration requirement of the crops and thus reduces net growth and productivity (Aggarwal, 2003). Pathak et al. (2003) reported the negative trends in solar radiation and an increase in minimum temperature, resulting in declining trends of potential yields of rice and wheat in the Indo-Gangetic plains of India.

Interception of radiation is not uniform in all parts of the crop. Distribution of light within the plant community is different from its distribution on a single plant or on a single leaf. Leaf orientation and leaf area index (LAI) are important factors for interception of light and radiation. Higher the LAI required intercepting the same fraction of light, the more the crop is efficient in utilising solar energy. The radiation profile also changes with height and day time hours. In general, interception is maximum in noon and minimum in morning and sunset. On an average, about 75% of incident radiation is absorbed by the canopy of crops, 15% is reflected and 10% is transmitted from the crop field. For the best utilisation of solar radiation, the lower 13% leaves should be oriented at an angle of 0-30^0, middle 37% at 30-60^0 and the upper 50% should be at 60-90^0 orientation. Actual utilisation of radiation by most crops is less than 1% and the efficiency varies from crop to crop. For example, the efficiency of utilisation in potato is 0.5%, 0.90% in wheat, 1.05% in maize and 1.43% in sugarcane. Usually the C_4 plants like sugarcane and maize are more efficient than C_3 plants like wheat and rice.

The ratio of photosynthetically active radiation (PAR) or simply light to total solar radiation is close to 0.50 in both tropics and temperate regions (Montheith, 1972) and represent a weighted mean between functions of direct radiation and different sky radiation.

There is a linear relationship between light intensity and photosynthesis for individual leaf. Net assimilation will be zero at light intensity of 500 ft. candle and a minimum of 500-1000 ft. candle will be required for effective rate of photosynthesis at which photosynthetic gain is greater than the respiratory loss (Blackman & Black, 1959).Extremely high light intensity may have an inhibitory effect on photosynthesis, a phenomenon called solarisation (Wang, 1963). Different plants require different hours of light for flowering. Long day plants (more than 14 hours), short day plants (less than 10 hours), and day neutral and indeterminate plants (requiring 12-14 hours) initiate flowering at different periods.

3.1.2 Water

Water is elixir of life. Plants require huge quantities of water which is required for photosynthesis, respiration, absorption, translocation and utilization of mineral nutrients, cell division and many other vital processes. It also affects evapotranspiration which is a joint process of evaporation of water from the Earth surface and plant surfaces and transpiration of water through the plant stems and tissues. Although plant hardly uses 1% of water absorbed for metabolic activities, evapotranspiration contributing 99% of water loss is a necessary evil. Water affects the performance of crops not only directly but also indirectly by influencing the availability of other nutrients, the timing of cultural operations etc. Water is a critical input which impacts crop yield. Excess water and water scarcity, both highly impact the quantity and quality of crop yield.

Although precipitation includes rainfall, snow, hail, fog and dew, rainfall is the main source of water which influences the crop growth, development and yield. Therefore, distribution of rainfall mainly decides the nature of cropping pattern in a place. However, the crop yield is not directly proportional to the amount of rainfall as uneven distribution or excess rainfall may reduce the crop yield. The rain that falls on the ground may be effective to the maximum extent. To be reasonably safe from evaporation, the rain must penetrate to a depth of at least 10-12 cm of soil (Staple, 1964). Annual effective rainfall is about one-third of the total rainfall (Bagnold, 1954). Rainfall analysis is required for decision making on sowing period, scheduling of irrigation, time of harvesting etc. Amount, intensity and distribution of rainfall are important for higher crop yields. Distribution of rainfall is more important than total rainfall. For example, 300 mm of rainfall falling on a day followed by rainless days for 2-3 weeks may create a drought situation whereas the same 300 mm evenly distributed in the critical month may double the crop yield. Dependability of rainfall is estimated by 75% of probability rainfall and by coefficient of variation. Length of growing season (LGS) also is rain dependent. It is the period between seed to seed when soil profile moisture is reduced to a ratio of AE/PE to 0.5, where AE and PE are actual and potential evaporation. Length of growing period during the year is number of days when the precipitation exceeds the 0.5 of PE plus a period required to evapo-transpire an assumed 100 mm from soil profile storage.

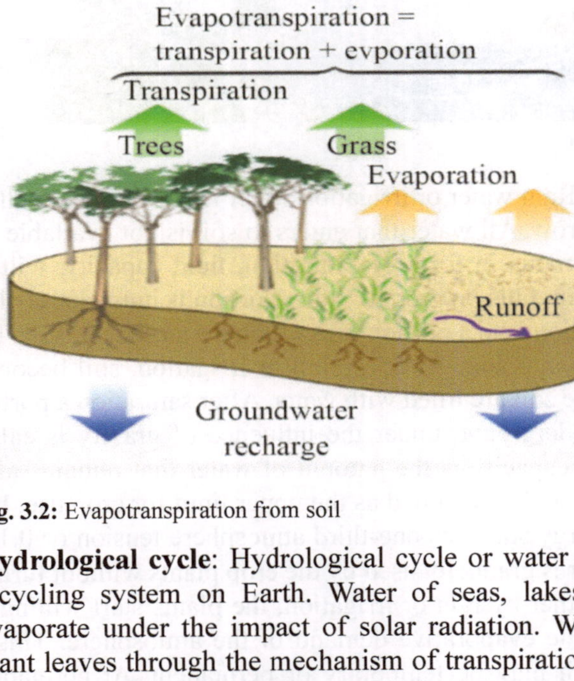

Fig. 3.2: Evapotranspiration from soil

Hydrological cycle: Hydrological cycle or water cycle is the normal water recycling system on Earth. Water of seas, lakes and other water bodies evaporate under the impact of solar radiation. Water also evaporates from plant leaves through the mechanism of transpiration. The water vapours cool in the atmosphere, condense as clouds and return to the land and the sea as precipitation. Precipitation falls on the Earth as surface water which flows as streams of water that result in lakes and rivers. A part of the water precipitating infiltrates into the ground and moves downward through the incisions, forming aquifers. Finally, a part of the surface and underground water moves to sea and remains as gas liquid and ice.

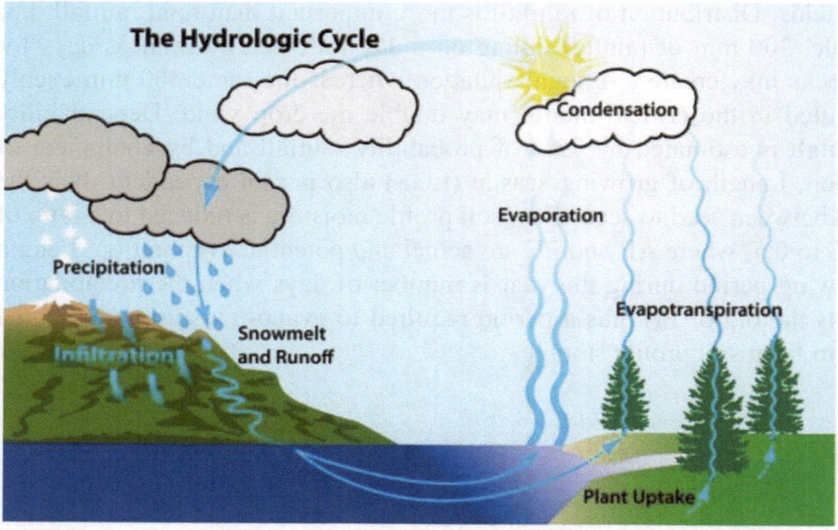

Fig. 3.3: Hydrological Cycle

Available soil moisture: Rain water or irrigation water is stored in soil which is required for plants to grow. All water that enters to soil is not available for plants. The soil water remains in form of saturation, field capacity, wilting point and vapour. When rain falls on ground, some portion is immediately lost to atmosphere by evaporation, and a portion runs off on the surface of soil. The remaining quantity enters into soil. After rainfall or irrigation, soil becomes saturated as all pores in the soil are filled with water. After saturation a portion of water drains to the underground under the influence of gravity is called gravitational water. Field capacity is the amount of water that remains after loss of gravitational water and is regarded as the upper limit for crop use. It is expressed in terms of energy status as one-third atmosphere tension or-0.1 to – 0.33 bar. Once soil water is gradually used by the crop plants without further replenishment of water either by rain or irrigation, the plants start wilting as they are unable to meet the evaporative demand of the atmosphere. This is called wilting point which may be temporary or permanent. At permanent wilting point the energy status of soil moisture reaches -15 bar. At this point plants wilt and survive as they are able to draw small amount of water from soil in cool hours. As the soil further dries, water is held in soil very tightly and the plants roots are not able to absorb water and start dying. The soil moisture at this stage reaches -60 bar and water is retained in vapour form. The limits of soil water between the field capacity and permanent wilting coefficient (FC-PWP) is the relatively available moisture (RAM) for the crops.

The water release characteristic curve is Indicated in Fig. 3.4 & 3.5.

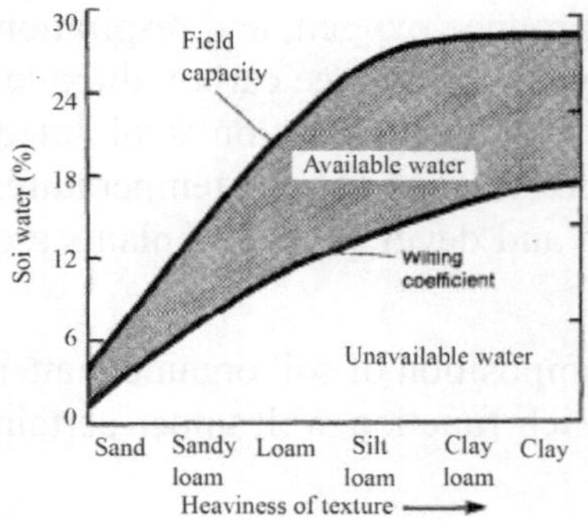

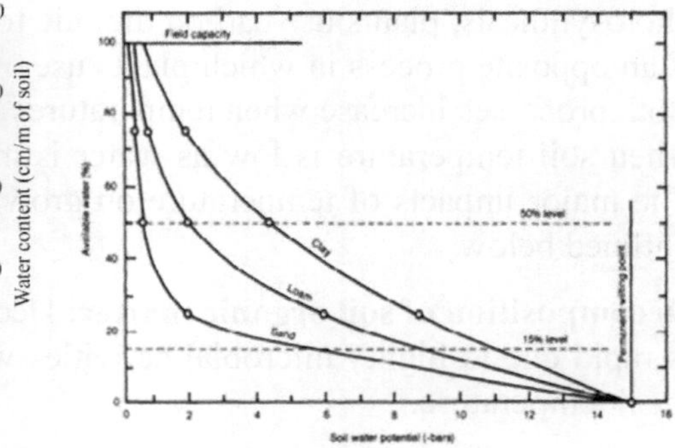

Fig. 3.4: Soil available water

Fig.3.5: Moisture extraction curve

Moisture extraction from different soil depths by roots of crops is not uniform. Plants extract maximum soil moisture from the upper zone. Approximately 40% of soil water is extracted from the upper zone followed by 30%, 20% and 10% respectively from second, third and fourth quarter zone (Fig.3.6)

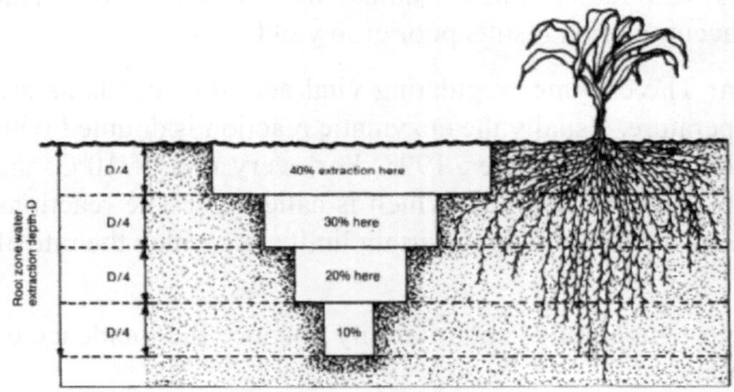

Fig. 3.6: Moisture extraction pattern from soil layers

3.1.3 Temperature

Temperature is an important factor limiting the distribution of plants and animals on a global scale. It influences evapotranspiration, photosynthesis, and soil warming in the crop field. Soil temperature is more important than atmospheric temperature with regard to seed germination, absorption of nutrients and root growth etc. while air temperature impacts all growth processes in plants including respiration, respiration and transpiration. Plants require certain range of temperature for all vital activities like respiration, photosynthesis and transpiration. A temperature range of 15°C to 40°C is required for most of the agricultural crops. High temperatures affect photosynthesis and respiration. In photosynthesis, plants use carbon dioxide to produce oxygen, and respiration is an opposite process in which plants use oxygen to produce carbon dioxide. Both processes increase when temperatures rise. Water absorption is inhibited when soil temperature is low as water is more viscous at cool temperature. The major impacts of temperature on growth and development of plants are outlined below.

Decomposition of soil organic matter: Decomposition of soil organic matter is rapid due to higher microbial activities which function well under certain ideal temperature.

Growth and development: At optimum temperature the activities of growth promoters like auxins, giberllins and cytokinnins and regulators like ascorbic acid are increased. Generally growth increases with rise in temperature up to a certain level after which it is reduced. However, requirement of optimum temperature depends on stages of growth.

Respiration: Respiration is high under increased temperature. High night temperature increases respiratory rates leading to exhaustion of plant assimilates. Poor net accumulation results poor crop yield.

Biochemical reaction: The enzymes regulating vital activities of plants are also regulated by temperature. Usually the enzymatic reaction is doubled with increase of temperature by 10°C (Mader, 1993).For every rise of 10°C, the rate of chemical reaction can be doubled, which is called Q_{10}. The reactions increase with increase in temperature up to certain limit after which the rate of reaction decreases.

Pest and disease: High temperatures and humidity enhance the incidence of most pathogens and pests.

Photosynthesis: Temperature regulates the formation of chloroplasts. At low temperature leaves become yellow due to degradation of chlorophyll. The rate of photosynthesis is reduced due to decrease in temperature. Recovery rate of photosynthesis is uniform in all leaves after the low temperature stress is relieved. But in case of higher temperature stress, only younger leaves can recover.

Temperature influences germination of seeds and different seeds have different requirements of optimal temperature for germination (Table 3.2).

Table 3.2: Effect of soil temperature on germination (%) of different crops

Crop	Soil temperature in 0°C and germination %								
	5	**10**	**15**	**20**	**25**	**30**	**35**	**40**	**45**
Rice	0	0	92	100	100	96	85	60	0
Wheat	40	97	97	100	100	97	85	18	0
Maize	0	0	93	100	100	96	85	60	0
Jowar	0	0	55	95	98	98	93	83	0
Cotton	0	0	18	93	96	98	98	83	0
Gram	35	98	100	100	100	100	100	55	0
Pea	38	98	100	98	98	63	0	0	0
Bottle gourd	0	0	0	44	97	97	97	53	0
Okra	0	0	0	73	100	98	100	85	0

Source: Singh, N.I and Dhaliwal, G.S. 1972, Plant and Soil, 37: pp 441-444

Temperature of soil is slightly more than the atmospheric temperature due to high specific heat of soil. Air temperature is maximum in summer (May) while it is the lowest in winter (Dec-January).The difference in maximum and minimum temperature is minimum during July to September due to more moisture in soil profile than the other months when the soil is dry. This indicates upward movement of heat from soil to atmosphere.

Plants grow less satisfactorily under constant temperature than when they are exposed to day-night differential temperature. This response is called thermoperiodism which varies with crops and crop stages. Low night temperature increases sugar content in many plants and translocation is favoured in temperature regime of 25^{0} C.

The temperature requirement of some common crop species is given in Table 3.2.

Table 3.2: Temperature requirement of some crops

Crop	Minimum temperature in 0°C	Optimal temperature in 0°C	Maximum temperature in 0°C
Rice	10	32	36-38
Wheat	4.5	20	30-32
Maize	8-10	20	40-43
Sorghum	12-13	25	40
Tobacco	12-14	29	35

Higher temperatures have more deleterious effect on net photosynthesis than lower temperatures which lead to decreased production of phosynthates above the critical temperature. Temperature often controls flowering on photoperiods. Information on accumulated day degree requirement for different cultivars allows sowing days to be planned to avoid concentration of sowing periods (Reddy, 1999).

3.1.4 Atmospheric Humidity

Relative humidity has direct influence on evapotranspiration and indirect effect on growth, photosynthesis, pollination, pest and disease infestation and yield. In dry regions the relative humidity remains low as a result of which there is water deficit in plants and reduced leaf water potential.

High relative humidity increases the incidence of crop disease and certain insect pests and reduces the ability of plants to intercept solar radiation. Low relative humidity increases evapotranspiration and transpiration causing wilting of leaves. The amount of water vapour in the air ranges from 0.01% by volume at the frigid poles to 5% in the humid tropics. The air becomes moist in high RH while it is dry in low RH. The RH affects the opening and closing of the stomata which regulates the loss of water from the plant through transpiration as well as photosynthesis. For plant propagation humid condition is more favourable. High humidity favours leaf enlargement while moderate RH favours pollination and seed setting. The water requirement of crop is also influenced by RH. Most of the crops perform better in RH of 40-60% while a few crops can perform well in RH of 80%.

3.1.5 Wind Velocity

Air in motion is called wind. Winds blowing over a warm current are usually laden with moisture while those blowing over a cold current have a cooling effect on the coast leading to formation of fog. Wind turbulence in atmosphere favours gas exchange.

The carbon dioxide and oxygen content of air is very important from photosynthesis and respiration point of view. Wind moving at a greater speed enhances the pace of evapotranspiration and reduces water balance in the leaves due to reduction of relative turgidity. When the rate of transpiration exceeds the rate of water absorption by roots, the leaf stomata may be closed and it will restrict the diffusion of carbon dioxide. This will cause decreased photosynthesis, growth and yield (Edmond et al, 1978). Wind over the crop surface can alter the onset of drought during dry periods due to water vapour loss through transpiration. This may lead to stomatal closure and reduced rate of gaseous exchange leading to reduced photosynthesis and crop growth. Moisture stress in the leaves causes anatomical and morphological modifications in plants due to limited soil moisture.

Wind can alter the balance of hormones in plants. Hot and dry wind cause reduction in plant height due to reduction in cell expansion. High velocity of wind may cause lodging of tall and weak plants. Mechanical damage may also be caused to leaf tips by strong wind.

The classification of wind is made as per its speed expressed in km/hour. The wind at speed of less than 1.5 km/hour is calm wind and 100-120 km/hour is called storm while at a speed exceeding 120 km/hour it is called hurricane. Gentle wind movement at a speed of 4-6 km per hour favours gas exchange and the wind speed at a higher rate can cause mechanical damage to the leaves and branches. Growth of plant is inhibited at wind above 10 km/h (Went, 1957). Favourable wind speed is essential for dispersal of pollens for seed setting. Excess speed of wind may cause soil erosion.

Monsoon is a flow pattern of general circulation in which there is a clearly dominant wind in one direction in every part of the region concerned. In India we experience SW and NE monsoon and the climate of the country is dominated by the SW monsoon winds which blow constantly from June to September. The heaviest rain falls in the north-eastern part of India, on the foothills of the Himalayan Mountains. Favourable monsoon is essential for higher crop production and aberration in the monsoon pattern causes low production of crops and food. While deficient rainfall causes drought, excess rainfall causes flood and inundation. In the winter season the monsoon changes its direction and blows as a dry cool wind from the inner parts of Asia over the Himalaya Mountains to India.

3.1.6 Atmospheric Gases

Most of dry air in the troposphere consists of 78.09% nitrogen and 20.95% oxygen with argon (0.93%), carbon dioxide (0.03%), and traces of hydrogen and other gases.

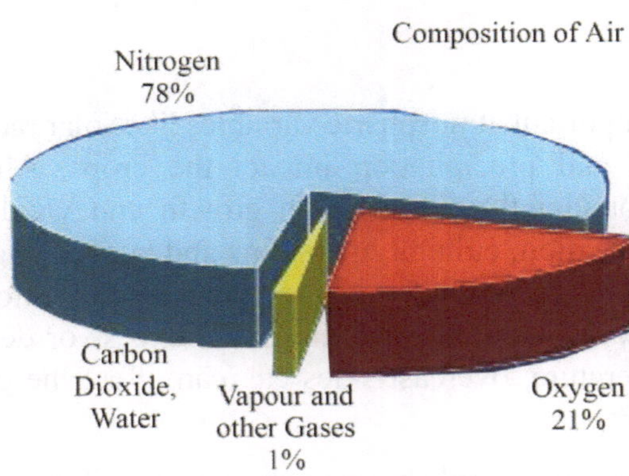

Fig. 3.7: Composition of air

Carbon dioxide is a raw material for photosynthesis in plants while is also a by-product of respiration. Oxygen is a by-product of photosynthesis and an input for respiration. Atmosphere can be polluted by suspended particles of dust and presence of carbon monoxide (CO), carbon dioxide (CO_2), sulphur dioxide (SO_2), sulphur trioxide (SO_3), nitrogen oxides, methane (CH_4), chlorofluorocarbon (CFC), and some suspended particles. Polluted air can influence the rate of photosynthesis and respiration, which ultimately affect the plant growth. Ozone depletion may lead to increase ultra violet radiation with adverse impact on Earth's environment and mankind (Paroda, 1998). Acid rain is often occurs due to high concentration of sulphur dioxide and sulphates.

Most plants growing in enhanced CO_2 concentration usually enhances net photosynthesis. The higher photosynthesis rates are then manifested in higher leaf area, dry matter production, and yield for many crops (Cure, 1985). CO_2 enrichment reduces transpiration per unit leaf area also. Higher CO_2 often improves water-use efficiency. Temperate crops are more efficient than tropical crops in utilisation of CO_2. Crops like rice, wheat, cotton etc. using C_3 pathway of photosynthesis are found to be less efficient in photosynthesis than C_4 crops like maize, sugarcane, sorghum and millets.

Nitrogen is one of the important major plant nutrients and gaseous nitrogen present in atmosphere cannot be used by plants directly. Atmospheric nitrogen is fixed in the soil by lightning and first rainfall. Biological nitrogen fixation by symbiotic association of Rhizobium bacteria found in root nodules of legume crops enriches the soil. However, non-symbiotic anaerobic bacteria such as Clostridium, Azotobacter and Azospirillum etc. are free living with nitrogen fixing potential. Both carbon cycle and nitrogen cycle are important biological cycles that operate in soil.

3.2 Weather and Crops

In addition to soil factors, important atmospheric variables like solar radiation, air temperature, humidity, and precipitation impact the crops. All these weather factors have profound influence on crop growth and yield. Field operations like nutrient application, cultural operations and pest management practices are also influenced by these weather factors. Weather aberrations like intermittent dry spell and early withdrawal of monsoon, excess or deficit of rainfall fluctuation in temperature, overcast skies etc. can affect the growth, development and yield of crops.

Dry soil may result poor stand establishment while wet soil may result loss of nitrogen through denitrification and leaching. This may lead to the development of seed, root, and crown diseases. The impacts on yield due to high temperature extremes have been modelled or empirically estimated at the global scale during the last years (e.g., Gourdji et al., 2013 and Deryng et al., 2014). Tropical cyclone may cause crop damage due to blowing tides, insufficient oxygen caused by overhead flooding, flash floods, wind injury to plant organs and water stress induced by enforced respiration, all of which occur at the same time (Iizumi et al., 2008 and Masutomi et al., 2012).

References

Aggarwal, P. K. (2003). Impact of climate change on Indian agriculture. J. Plant Biology 30:189–198.

Bagnold, R.A, 1954. The physiological aspects of deserts. In: J.L. Coundsley-Thompson (ed). Biology of deserts, Institute of Biology, London.

Blackman, G.E & Black, J.N, 1959: Physiological and ecological studies in analysing plant environment. XII.The role of light as limiting factor. Anals of Botany, 23: 131-145

Challinor, A., Watson, J., Lobell, D. et al. A meta-analysis of crop yield under climate change and adaptation. Nature Clim Change 4, 287–291 (2014). https://doi.org/10.1038/nclimate2153

Cure, J.D. 1985. Carbon dioxide doubling response: A crop survey. In: Direct Effects of Increasing Carbon Dioxide on Vegetation. (B.R. Strain and J.D. Cure (Eds.)). DOE/ER-0238, US Dept. of Energy, Washington D.C., 99-116.

Delvin, R.M, 1975: Plant physiology

Deryng D, Sacks W J, Barford C C and Ramankutty N 2011 Simulating the effects of climate and agricultural management practices on global crop yield Glob Biochem. Cycles 25 GB2006

Deryng, D et al, 2014: Global crop yield response to extreme heat stress under multiple climate change futures: Published 19 March 2014 : IOP Publishing Ltd Environmental Research Letters, Volume 9, Number 3

Doorenbos, J and Pruitt, W O 1984 Guidelines for predicting crop water requirements FAO Irrigation and Drainage Paper 24 (Rome: FAO)

Eagleman, J.R.1985, Meteorology: The Atmosphere in Action, Wadsworth Publishing Company. 394 p

Eaglman, 1985:Metereology: The Atmosphere in action: environmental factors that affect plant growth.

Gourdji S M et al. 2013 Global crop exposure to critical high temperatures in the reproductive period: historical trends and future projections Environ. Res. Lett. 8 024041

Gourdji S .M., Mathews, K. L., Reynolds, M, Crossa, J and Lobell, D. B 2013: An assessment of wheat yield sensitivity and breeding gains in hot environments Proc. R. Soc. B 280 20122190

Grace, J, 1977. Plant response to wind. Academic Press, London

Koide, N. et al 2013: Prediction of Rice Production in the Philippines Using Seasonal Climate Forecasts: Journal of Applied Meteorology and Climatology 52(3):552-569

Mader S.S.1993, Biology Part 1: The Cell. 4th ed. Duburque, IA, USA: Wm. C. Brown Communications, Inc. 152 p

Masutomi, Y. et al 2012: Estimation of the damage area due to tropical cyclones using fragility curves for paddy rice in Japan :Environ. Res. Lett. 7 (2012) 014020 (9pp) doi:10.1088/1748-9326/7/1/014020

Miller, C.A and Edwards, P.N, 2001: Changing the Atmosphere: Expert Knowledge and Environmental Governance

Montheith, J.L.1972: Solar radiation and productivity of tropical ecosystem. Journal of Applied Ecology, 9: 747-766

Paroda, R.S, 1998. Natural resource management for sustainable agriculture: A new paradigm. Plenary lecture at first international agronomic congress, 24 Nov, 1993, ND.

Pathak, H., Ladha, Aggarwal, P. K., Peng, S., Das, S., Singh, Y., Singh, B., Kamra, S. K., Mishra, B., Sastri, A. S. R. A. S., Aggarwal, H. P., Das, D. K. and Gupta R. K.(2003). Trends of climatic potential and on-farm yields of rice and wheat in the Indo-Gangetic Plains. Field Crops Res., 80: 223–234.

Porter, J .R and Gawith, M 1999 Temperatures and the growth and development of wheat: a review Eur. J. Agronomy 10 23–36

Porter, J.R., Xie, L., Challinor, A.J., Cochrane, K., Howden, M., Iqbal, M.M., Lobell, D.B.,Travasso, M.I., 2014. Chapter 7. Food Security and Food Production Systems. Climate Change 2014: Impacts, Adaptation and Vulnerability. Working Group IIContribution to the IPCC 5th Assessment Report, Geneva, Switzerland

Reddy, S.R.1995: Principles of Agronomy, Kalyani Publishers, Ludhiana, India.

Sakamoto, M. & Benno, Y. (2006). Reclassification of Bacteroides distasonis, Bacteroides goldsteinii and Bacteroides merdae as Parabacteroides distasonis gen. nov., comb. nov., Parabacteroides goldsteinii comb. nov. and Parabacteroides merdae comb. nov. Int J Syst Evol Microbiol 56, 1599–1605

Shantz, HL and Lydia NP, 1925: The water requirement of plants: J. Ag Res Vol 34(12), Washington DC, 1927

Staple, W.J, 1964. Dryland Agriculture and Water Conservation. In: Research on water, ASA Special publication. Soil Science Society of America, Madison, Wisconsin.

Wang, J.Y, 1963. Agricultural Meteorology. Pacemakers Press, Milwaukee, Wisconsin.

Went, F.W. 1957. Experimental control of plant growth, chronica Botanica 17, Ronald Press Co. New York.

4

Impacts of Climate Change on Agriculture

4.1 Climate and Agriculture

Soil, climate and water are three basic requirements of a crop to be grown in a particular region. Climate affects the temperature trends, seasons, length of the growing season, amount of sunlight and the seasonality of sunlight etc. A change in climate may cause recurrent droughts, floods and weather extremities that would affect the agricultural production, human livelihoods, health and biodiversity. The relationship between agriculture and climate is complex as it is dependent on climate and also plays a contributing role in emission of greenhouse gas (GHG). Greenhouse gases include carbon dioxide, nitrous oxide, methane, chlorofluorocarbons (CFCs), hydro-chlorofluorocarbons (HCFCs), hydro-fluorocarbons (HFCs), per fluorocarbons (PFCs) etc. , which play a significant role in climate change.

Climate change is experienced by all and everyone feels its effects. The farmers are already experiencing the adverse effects of climate change. There are many deviations in weather patterns. Temperature increase is 1.1 degrees Celsius above the pre-industrial level due to emissions of GHG. When these gases are emitted into the atmosphere, they trap heat close to the earth's surface due to greenhouse effect. This results in global warming, which is an important cause of climate change. The effect of climate change is seen in forms of more intense rainfalls, longer dry spells, and natural disasters. Agriculture itself is also a cause of greenhouse gas emissions contributing to climate change. Direct emissions in form of methane from animals and carbon dioxide from soils have resulted in increased quantities of these gases being emitted to the atmosphere. The faulty methods of cultivation and livestock management contribute to emission of greenhouse gasses. The extreme weather reduces crop production and livestock system including fisheries, which affects their livelihoods and food security.

4.2 Factors Contributing to Climate Change

Climate change is caused due to (i) geo-ecological events and (ii) human activities.

4.2.1 Geo-ecological Events

Occurrence of earthquakes, tsunami, global warming/cooling, global dimming, and incursions of sea water are major geo-ecological events causing climate change.

Continental drift: The continental drift has an impact on the climate because it changed the physical features of the landmass, their position and the position of water bodies. The separation of the landmasses changed the flow of ocean currents and winds, which affected the climate.

Volcanoes: When a volcano erupts, it throws out sulphur dioxide (SO_2), water vapour, dust, and ash into the atmosphere. Volcanic activity lasts for a few days, but the amount of gases and ash it releases can influence climatic patterns for years. Large quantity of sulphur dioxide gas can reach the stratosphere from a major eruption. These gases and dust particles partially block the incoming rays of the sun, leading to cooling. Sulphur dioxide combines with water to form tiny droplets of sulphuric acid. These droplets are so small that many of them can stay in the sky for several years. They are efficient reflectors of sunlight and screen the ground from some solar energy.

The Earth's tilt: The earth makes one full orbit around the Sun each year. It is tilted at an angle of 23.5° to the perpendicular plane of its orbital path. For one-half of the year when it is summer, the northern hemisphere tilts towards the Sun. In the other half when it is winter, the Earth is tilted away from the sun. Such tilt of Earth causes seasons like summer and winter. More tilt causes warmer summers and colder winters while less tilt causes cooler summers and milder winters. This gradual change in the direction of the Earth's axis is responsible for changes in the climate.

Ocean currents: The oceans cover about 71% of the Earth and absorb about twice as much of the Sun's radiation as the atmosphere or the land surface. Ocean currents move vast amounts of heat across the planet which is approximately same as the atmosphere. As the oceans are surrounded by land masses, heat is transported through channels. Winds push horizontally against the sea surface and drive ocean current patterns. The El Niño event in the Pacific Ocean can affect climatic conditions of the world. Ocean currents may change direction or slow down. Much of the heat that escapes from the oceans is in the form of water vapour. The water vapour is also the most abundant greenhouse gas on Earth and plays a role in climate change. However, water vapour also contributes to the formation of clouds.

Earthquakes: Earthquake is caused naturally and by human activities. Sudden release of stored energy in the Earth's crust creating seismic waves results earthquake, which is manifested by sudden shaking or displacement of the ground. An earthquake caused by tectonic plates put a strain on the ground which becomes so great that rocks give way breaking and sliding along faulty planes. Sometimes earthquakes cause tsunamis that may cause loss of property and lives. Smaller earthquakes can be caused by volcanic activity, landslides, mine blasts, and nuclear experiments.

According to new research, earthquake could contribute towards climate change by releasing methane that is held in pockets within the Earth. These pockets are generally located at sub-sea level and can be released by the shifting of tectonic plates. The study performed by German and Swiss scientists has suggested that methane can be released from sub-sea pockets during earthquakes. According to many studies, there is a correlation between global temperature and magnitude of earthquake. Temperature raises heat fluxes that may enhance evapotranspiration and precipitation. Surface latent heat flux, rainfall, cloud coverage, vertical air motion, and wind speed near the surface are the key climatic variables.

Tsunami: With the rapid displacement of water from ocean a series of waves are created which is called, tsunami. The factors responsible for generation of tsunami are earthquakes, volcanic eruption, mass movement of water from ocean, explosions and landslides etc. Post tsunami impacts are seen in the form of littering the paddy fields with debris and sediments, destruction of irrigation canals, development of salinity which ultimately affect the agriculture production and livelihoods of farmers and fishermen.

Global warming: The increase of average Earth's temperature is called global warming. Over the last 100 years, there has been rise in temperature of air near the Earth's surface by about 1^0C. Since the Earth is heated by this temperature rise, many disasters like hurricanes, droughts and floods have frequently occurred. The human actions like emission of greenhouse gasses, burning of forests and deforestation etc. are responsible for global warming.

Global cooling: Global cooling was the result of extensive glaciation and the cooling effects of aerosols or orbital forcing. Paul R. Ehrlich (1968) mentioned global warming from greenhouse gases as a counterforce to the cooling effect of aerosols . However, the probability of this occurring is generally considered to be very low.

There is now general agreement that aerosol effects were the dominant cause of the mid-20th century cooling. Due to fossil fuel combustion, tiny dust particles, called aerosol, are increased in atmosphere. Aerosols directly increase the planetary albedo, which reduces the solar radiation from reaching the surface and causes cooling. Indirectly they affect the properties of clouds by acting as cloud condensation nuclei. (Rasool &Schneider, (1971). Orbital forcing is slow, cyclical changes in the tilt of Earth's axis and shape of its orbit, which affect the timing and intensity of the seasons and change the amount of sunlight reaching the Earth. In comparison to global warming the risk of global cooling is very small. However, a sudden cooling may affect the crops and release of cold-tolerant crop varieties would be necessary to avoid major losses.

Global dimming: Gradual reduction of irradiance at the Earth's surface is called global dimming. Increase in particulates such as sulphate aerosols in the atmosphere is believed to be reason for such global dimming. It may interfere with the hydrological cycle by reducing evapotranspiration which causes reduced rainfall in some places.

4.2.2 Human Factors

Among artificial factors industrial revolution and release of GHG to the atmosphere are the major causes of climate change.

4.2.2.1 Industrial Development

Industrialization is very imperative for economic growth of a nation. In spite of improving living standards of people in certain aspects, it impacts the environment in way of atmospheric pollution and contributes to climate change._Industrialisation without precautionary measures is known to cause atmospheric pollution. The greenhouse gasses formed in the atmosphere are water vapours, carbon dioxide, methane, nitrous oxide and CFC etc.

- Water vapour (H_2O) is gaseous form of water that evaporates from the sea, lakes, rivers and the soil surface, and that is transpired by plants and often felt as humidity. Human activity makes little direct contribution to the large amount of water vapour or clouds in the atmosphere. Water vapour increases as the Earth's atmosphere warms and precipitates on condensation.
- Carbon dioxide (CO_2) can be found in nature produced by volcanoes and geysers. Transport and energy production based on combustion engines burning fossil fuels such as coal, mineral oil and gas emit carbon dioxide and carbon monoxide to the atmosphere. All animals exhale it through respiration. Plants also exhale carbon dioxide during respiration at night in absence of photosynthesis. It also enters the atmosphere through the decay of organic matter, deforestation, burning vegetation and certain industrial processes such as cement-making. Humans have increased atmospheric CO_2 concentration by 48% since the Industrial Revolution began which is a major cause of climate change.

- Methane (CH_4) enters the atmosphere when produced by livestock, as well as by microbes in the soil and in flooded water of rice fields. This hydrocarbon gas is produced both through natural sources and human activities, including the decomposition of wastes in landfills, agriculture, and especially rice cultivation, as well as ruminant digestion and manure management in livestock production. Methane is more active greenhouse gas than carbon dioxide, but also one which is much less abundant in the atmosphere.
- Nitrous oxide (N_2O) is a powerful greenhouse gas produced by cultivation practices in the crop fields, especially by the use of commercial and organic fertilizers, fossil fuel combustion for mechanisation and stubble burning.
- Chlorofluorocarbons (CFCs): Fluorinated gases (F-gases) are made by humans and used in refrigerators, air-conditioners, foams, cosmetics and fire extinguishers. They act as pollutants and play a role in climate change.

The causes of emission of GHGs have been shown in Fig.4.1. The greenhouse gasses accumulate in the atmosphere as a result of burning fossil fuels. Such gasses act like a blanket around the Earth and trap too much of heat that would have escaped into space. The key challenges include coping with higher temperatures, extreme weather conditions affecting human life, livestock and agriculture.

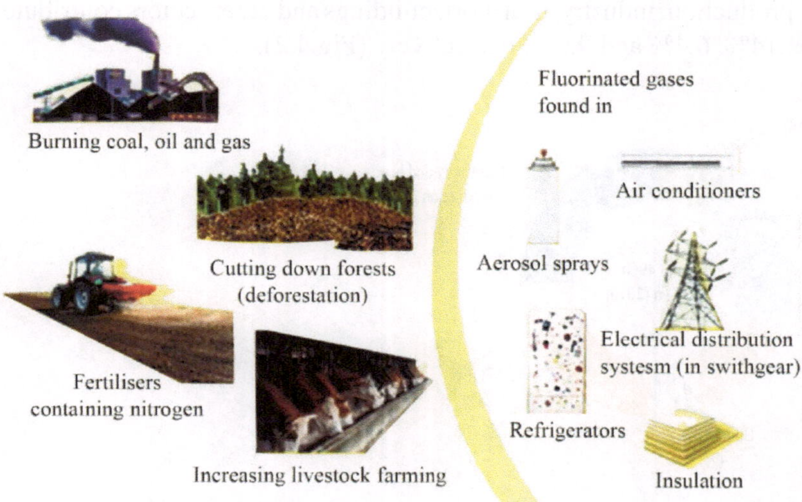

Fig. 4.1: Causes of greenhouse gas emission

Along with industrialisation the following activities have also been responsible for climate change.

Power generation: Power generation by burning fossil fuels such as coal, gas, and gas produce GHG such as carbon dioxide and nitrous oxide that blanket the Earth and trap the Sun's heat. This can be reduced by using wind, solar and other sources of renewable energy.

Manufacturing goods: Production of cement, iron, steel, electronics, plastics, clothes, and other goods including mining and other industrial manufacturing release huge amount of GHG to the atmosphere as the machines used in the manufacturing process use fuels like coal, oil or gas.

Transportation: In transportation, the vehicles like cars, trucks, ships, and planes run on fossil fuels and contribute appreciably to GHG emissions. Road vehicles use mostly petrol, diesel and gasoline in internal combustion engines and emit major quantity of GHG to the atmosphere.

Powering buildings: The electricity produced from coal, oil or natural gas is used in buildings for heating and cooling. Thus they emit significant quantities of greenhouse gas. Increasing demand for air conditioners, fridges and lighting contribute to a rise of energy-related carbon-dioxide emissions.

4.2.2.2 Intensive Agriculture

Agriculture, forestry and other land use (AFOLU) sectors are the second most important source of emissions contributing around 24 per cent whereas electricity production, industry, transport, buildings and other sectors contribute 25%, 21%, 14%, 6.4% and 9.6%, respectively (Fig.4.2).

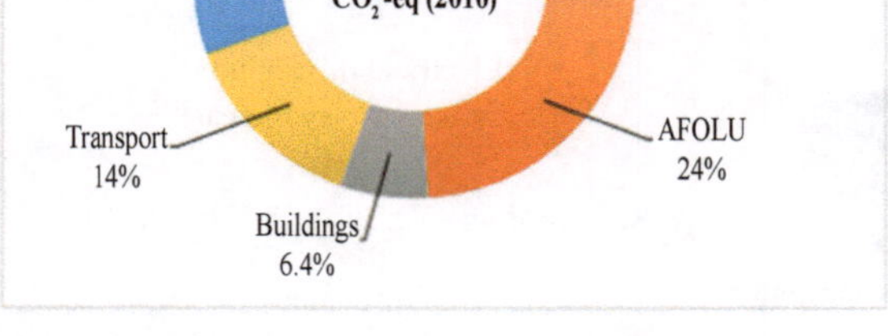

Fig. 4.2: Global greenhouse emission by economic sector
Source: IPCC, 2014 IPCC, 2014a

The greenhouse gases of particular relevance to these sectors include carbon dioxide (CO_2), methane (CH_4) and nitrous oxide (N_2O). Emissions of GHG from agriculture and allied sectors excluding forestry comes from enteric fermentation in livestock, leftover manures in the field, manure management, rice cultivation in wetland, burning of stubbles and crop residues and use of synthetic fertilisers etc. (Fig.4.3).

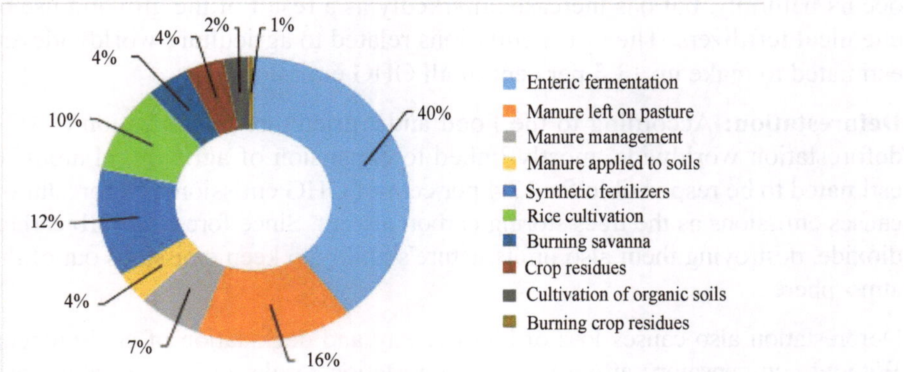

Fig. 4.3: Greenhouse gas emission by agriculture
Source: FAO, 2016b Annex table A.3

Soil emissions: Nitrous oxide emissions account for about 60 per cent of total agricultural sector emissions which is produced naturally in soils through the microbial processes of nitrification and de-nitrification. Excess use of nitrogen fertilizer for higher crop production has also increased emissions.

Enteric fermentation: During digestion, microbes in the animal's digestive system ferment feed. This process of enteric fermentation produces methane as a by-product which can be emitted by the exhaling and belching of the animal. Large ruminants have higher methane emissions than pigs and poultry because of their unique digestive systems.

Manure management: Methane is also produced by the anaerobic decomposition of manure. When manure is handled as a solid or deposited naturally on grassland, it decomposes aerobically and creates methane emissions.

CO_2 from fossil fuel consumption: Carbon dioxide emissions are indirectly caused primarily from combustion of gasoline and diesel to fuel farm equipment, including tractors, combines, irrigation pumps, grain dryers, etc. It also includes emissions related to the production of fertilizers, pesticides and herbicides, which are primarily derived from fossil fuels.

Rice cultivation: Rice is water guzzling crop and widely cultivated under submergence. Flooding causes decomposition of manure, soils and other organic matter on the fields in an anaerobic environment, and decomposition of these materials result in emission of methane that is released into the atmosphere.

Fertilisation: A relatively new greenhouse gas threat is nitrous oxide, which occurs naturally, but has increased markedly as a result of the growing use of chemical fertilizers. The direct emissions related to agriculture worldwide are estimated to make up 13.5 per cent of all GHG emissions.

Deforestation: According to the Food and Agriculture Organization (FAO), deforestation worldwide, mostly linked to expansion of agricultural areas, is estimated to be responsible for 17.4 per cent of GHG emissions. Deforestation causes emissions as the trees storing carbon are cut. Since forests absorb carbon dioxide, destroying them also limits nature's ability to keep emissions out of the atmosphere.

Deforestation also causes loss of bio-diversity and degradation of environment. Without compensatory afforestation, deforestation results in loss of habitat and biodiversity, reduction of wood and other raw materials for industry. It also affects rainfall pattern in the region.

Water storage in dams: Reservoirs storing water for irrigation may influence the micro-climate of the surrounding area in addition to causing waterlogging and salinity in adjacent areas.

4.2.2.3 Greenhouse Effect

The energy coming from the sun as ultraviolet radiation drives the Earth's weather and climate. Most of that energy is radiated back into space, but some is trapped in the ground, the ocean and the atmosphere. Nitrogen, oxygen and argon are major gases of atmosphere which have no such direct effect on climate. But the greenhouse gasses act like the glass in a greenhouse and block the heat for which the atmosphere warms up. This is known as the greenhouse effect. The Earth's atmosphere has always acted like a greenhouse to capture the Sun's heat and influences the life forms. Global warming is a major cause of climate change. The greenhouse effect is displayed in Fig.4.4.

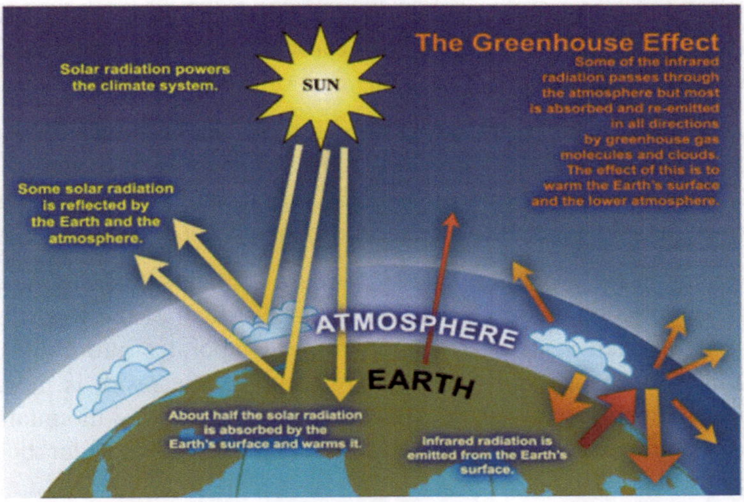

Fig. 4.4: The greenhouse effect

Carbon dioxide equivalents

Different greenhouse gases have different effects depending on their concentration in the atmosphere, ability to alter radiation balance and time of availability. To make comparisons of the global warming impacts of different gases, the IPCC (IPCC, 2014a) uses global warming potential (GWP) as a measure to "represent the combined effect of the differing times these substances remain in the atmosphere and their effectiveness in causing radiative forcing". The larger the GWP value, the higher is the warming effect of a given gas compared to carbon dioxide over a time period, usually in terms of 100 years. For example, one tonne of methane is equivalent to 25 tonnes of carbon dioxide, while one tonne of nitrous oxide is equivalent to 298 tonnes of carbon dioxide (IPCC, 2007b).

- A molecule of carbon dioxide has the smallest effect on the climate, but it is by far the most common anthropogenic greenhouse gas, so it has the biggest overall effect. The GWP of carbon dioxide is 1 regardless of the time period and it is used as the reference that allows comparison with other gases. Carbon dioxide accounts for around three-fourths of total emissions and its overall contribution to radiative forcing continues to rise. It is difficult to provide a single lifetime of carbon dioxide because the gas is not destroyed over time, rather it moves through the Earth system. Most of its excess has been incorporated in the oceans over decades, while the rest will remain in the atmosphere for thousands of years.

- Methane is a more potent greenhouse gas than carbon dioxide, but sunlight converts methane molecules to carbon dioxide after about 12 years. Averaged out, the GWP over 100 years of methane is 28.
- Other gases are more powerful sources of warming and last much longer in the atmosphere. Nitrous oxide lasts 121 years and has a GWP over 100 years of 265, while some types of fluorinated gases may last thousands of years with a GWP over 100 years of more than 6500.

Because all systems in the global climate system are connected, adding heat energy causes the global climate as a whole to change. Much of the world is covered with ocean which heats up and more water evaporates into clouds. Where storms like hurricanes and typhoons are forming, the result is more energy-intensive storms. A warmer atmosphere makes glaciers and mountain snow packs, the Polar ice cap, and the great ice shield jutting off of Antarctica melt raising sea levels.

Main sources of GHG

Human activities play an important role in the emission of greenhouse gases. The amount of carbon dioxide released by each country depends on the type and share of its main economic activities. Therefore, the shares of carbon dioxide emissions are different across the regions of the world. The latest estimates of emissions from human activities totalled more than 46 billion metric tonnes of greenhouse gases, expressed as carbon dioxide equivalents, representing a 35 per cent increase over the previous decade. Electricity and heat production represents the biggest source of greenhouse gases emissions globally, mainly due to the burning of coal, oil and natural gas (Figure 4.3). Agriculture, forestry and other land use (AFOLU) sectors are the second most important source of emissions, responsible globally for around 24 per cent. Other sectors include industry that accounts for 21 per cent of all emissions, transport – road vehicles, trains, ships and aircraft running on fossil fuels – accounting for 14 per cent, other energy sources accounting for 9.6 per cent and buildings accounting for 6.4 per cent (Fig.4.2).

4.3 Impacts of Climate Change

Climate change affects plants and animals by changing their natural equilibrium inside directly or indirectly. The loss of forest cover may cause soil erosion, floods and droughts, because the trees normally intercepts rain water and encourages infiltration. Lack of trees also exacerbates drought in dry years by making the soil dry more quickly due to surface evaporation. The global warming due to emission of greenhouse gases increases global atmospheric temperature over a long period of time. Such changes in surface air temperature and consequent adverse impact on rainfall over a long period of time are known as climate change. If these parameters show year-to-year variations or cyclic trends, it is known as climate variability. However, the official definition by the United Nations Framework Convention on Climate Change (UNFCCC) is that "Climate change is the change that can be attributed directly or indirectly to human activity that alters the composition of the global atmosphere and which is in addition to natural climate variability observed over comparable time periods". However, scientists often use the term for any change in the climate, whether arising naturally or from human causes.

4.3.1 Impact at Global Level

The likely impact of climate change at global level is explained below and summarised in Table 4.1.

Water: Both drought and flood will occur and the water availability is likely to decrease in the dry tropics and in mid-altitudes due to climate change. Melting of water in mountain ranges may decrease.

Food: The increased temperature and frequency of droughts and floods due to climate change may reduce the crop production and availability of food, especially in dry and tropical regions. It may affect all four dimensions of food security (food availability, food access, food use and sustainability). Climate change can disrupt food availability, reduce access to food and affect food quality and sustainability. Higher CO_2 levels can increase plant growth particularly in plants with C_3 photosynthetic mechanism (rice, oilseeds, pulses, vegetables etc.). Increased CO_2 level also reduces the quality of produce by reducing protein, nitrogen and minerals content in most plant species, including wheat, soybeans and rice that can threaten to human health. Climate change also affects weeds, pests and fungi infestation due to an increase in host susceptibility. The ranges and distribution of weeds and pests are likely to increase with climate change. This could cause new problems for farmers' crops previously unexposed to these species. Both historical and future estimated impacts of climate change on cereal crop yields in different locations indicate that the yield loss can be up to −35% for rice, −20% for wheat, −50% for sorghum, −60% for maize and −13% for barley depending on the location, future climate scenarios and projected year.

Industry, settlement and society: The industries located in coastal and river flood plains will be more affected by weather extremes. Due to more intense weather extreme events the economic and social costs will increase.

Health: Heat waves, floods, cyclones, storms, fires and droughts caused by climate change may affect the health of people causing increased deaths and diseases. The problems of malnutrition, diarrhoea disease and malaria in some areas will increase.

Warming and sea level rise: It is projected that throughout the Twenty first century, India will experience warming above global level. It is also expected that India will experience more seasonal variation in temperature with more warming in the winters than in the summers. In recent years, extreme heat waves across India have extended together with warmer night temperatures and hotter days, and this trend is expected to continue. This is predicted that the average temperature will rise up to 2.33–4.78°C with a doubling in CO_2 concentrations at the end of this century. These heat waves will lead to increased variability in summer monsoon precipitation, which will result in drastic effects on the agriculture sector in India.

Natural calamities: Due to rainfall variability, changes in the frequency and severity of droughts and floods could pose challenges for farmers and ranchers and will ultimately threaten global food security. Meanwhile, the rise in sea level and warming of sea water are likely to cause the habitat ranges of many fish species to shift, which could disrupt ecosystems.

Overall, climate change would adversely affect the crops, animals and marine life. Along with effects of climate change on agriculture, other evolving factors that affect agricultural production, such as changes in farming practices and technology, need to be addressed.

Table 4.1: Effects of climate change on several aspects of human life and ecosystem

Ocean	**Sea level rise:** Melting ice sheets and glaciers are responsible for rise in sea levels that inundate low-lying areas during high tides and storms. Islands, coastal cities and farmland near the sea are more vulnerable. **Acidification:** Dissolved carbon dioxide makes seawater more acidic, harming marine life. **Marine oxygen concentration:** Less dissolved oxygen in sea water affects marine life and can lead to changes in the distribution of species. **Increase temperature:** The types of plants, corals and animals that can live in each location will change and aquatic species will migrate due to warmer water temperatures.

Natural Resources	**Freshwater:** Availability of freshwater for drinking, domestic use and irrigation may decline. Rivers, springs and wells may dry up as air temperatures increase. **Vegetation:** Agro-ecological zones may shift and tree and plant ranges may move pole ward and up mountain slopes. **Wildlife:** The areas in which animals live may change and some animals and plants may move towards cooler places or higher altitudes. The areas of some species may expand while others will shrink. **Biodiversity:** The diversity of animals and plants is likely to decline, which will affect the stability of ecosystems, forests, grasslands and deserts. The area suited to forests and grasslands may change. In some areas, deserts may expand. Risks of forest fires may increase. **Extinctions:** Up to one-quarter of the world's species may become extinct as their habitats decline. **Ice sheets:** On-going melting of the ice sheets will occur in Antarctica and Greenland. **Glaciers:** They may recede and disappear and disrupt the quantity and timing of river flows.
Human Health	**Disease:** Some types of diseases, such as malaria, diarrhoea and dengue fever, may become more widespread. **Heat waves**: Periods of extreme high temperature may increase in intensity and frequency putting increasing numbers of people at risk.
Society & Economy	**Migration:** Changes may force people to move from one place to another in search of livelihoods within countries or across international boundaries. Those who have moved to coastal areas will likely be displaced again due to rise in sea level. **Conflict**: Competition for resources such as water, food and fuel may cause conflicts. **Poverty**: Some people are less able to adjust than others under climate change. Poverty levels are likely to increase. **Economy:** Climate change will affect businesses and the economy. It could cost between 5 and 20 per cent of the annual gross domestic product.

Sources: FAO, 2016a; IPCC, 2014b

4.3.2 Projected Impact on Asia

South Asia may be worst affected by climate change and food security of nearly 1.6 billion people will be threatened. The crop model predictions indicate that the yield level of rice and wheat would be reduced by 17% and 50% in 2050 from the 2000 level due to climate change. It is also projected that the availability of average calories in 2050 is likely to be reduced by 15%. It will increase the number of malnourished children.

4.3.3 Observed Changes in Climate and Weather Events in India

The surface temperature has increased by 0.4 C at the national level during the past century. Increased monsoon rainfall (+10% to +12% of the normal over the last 100 years) along west coast, northern AP and north-west India whereas decreasing trend has been observed over eastern Madhya Pradesh, north-eastern India, and some parts of Gujarat and Kerala (-6% to - 8% of the normal over the last 100 years) .

Extreme weather trends have been observed in many parts of the country. Scientists have estimated that sea level rise was between 1.06-1.75 mm per year. The available monitoring data on Himalayan glaciers indicates that the trend is not consistent across the entire mountain chain.

4.3.4 Some Projections of Climate Change over India for the 21st Century

The simulation studies by Indian Institute of Tropical Meteorology (IITM), Pune, estimated that annual mean surface temperature is expected to rise by the end of century, ranges from 3 to 5° C with warming more pronounced in the northern parts of India. Climate change may have significant long-term impact on quality and quantity of water. River systems of the Brahmaputra, the Ganga, and the Indus, which benefit from melting snow in the lean season, are likely to be particularly affected by the decrease in snow cover. A decline in total run-off for all river basins, except Narmada and Tapti, is projected. Due to sea level rise, the fresh water sources near the coastal regions will suffer salt intrusion.

4.4 Impacts of Climate Change on Agriculture

Agriculture is a climate dependent biological production system. Climate change disturbs the agricultural ecosystem with changes in agricultural climatic elements such as temperature, precipitation and sunlight. The direct solar radiation heating the Earth's surface is trapped by greenhouse gasses and water vapours, which causes global warming. The causes and effects of climate are displayed in Fig. 4.5.

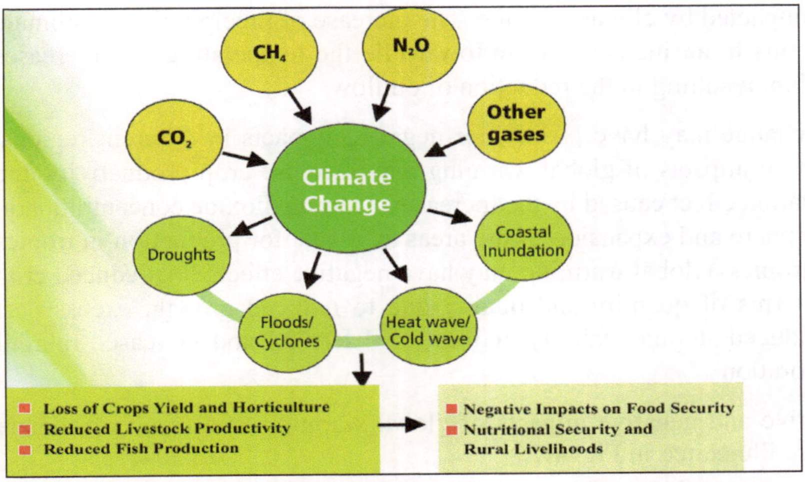

Fig. 4.5: Cause and impacts of climate change on agriculture and allied sectors

Changes in the climatic elements will affect the agricultural and livestock production. The flow of impact of climate change on agriculture sector has been illustrated in Fig. 4.6.

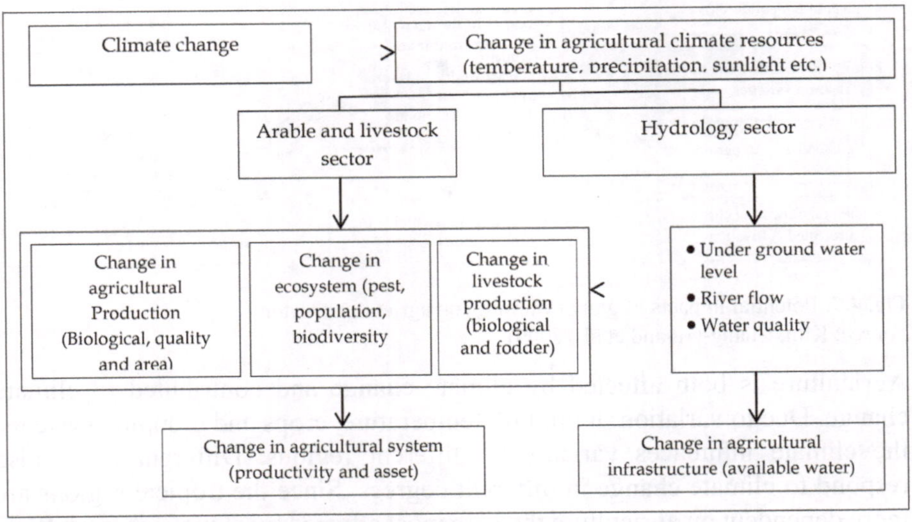

Fig. 4.6: Flow of the climate change impact on agriculture
Source: Kim,Chang-Gil et al (2009)

The hydrology including underground water level, water temperature, river flow, and water quality of lakes and marshes are impacted by climate change. Atmospheric temperature, precipitation, evaporation, and soil moisture content are also impacted by climate change. The increase of precipitation by climate change leads to an increase of outflow while the temperature rise increases evaporation, resulting in the reduction of outflow.

Climate change may have positive or negative impacts in different regions. The positive impacts of global warming may increase crop productivity due to fertilization effect caused by the increase in carbon dioxide concentration in the atmosphere and expansion of the areas available for production in tropics and sub-tropics. Global warming may have negative effects like reduced crop yield in terms of quantity and quality due to reduced growth, excess pest attack, reduced storage stability, reduced soil fertility and increased rainfall and evaporation.

The positive and negative impacts of global warming as a result of climate change are illustrated in Fig. 4.7.

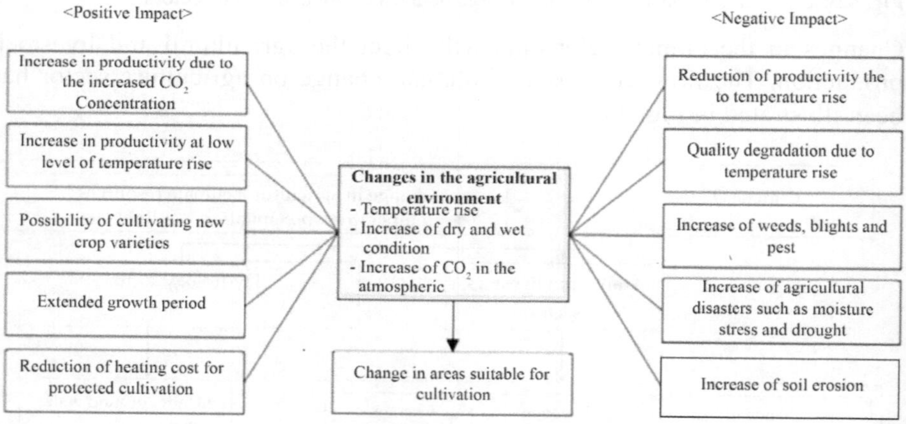

Fig. 4.7: Potential impacts of global warming on agriculture sector
Source: Kim, Chang-Gil and et al. (2009)

Agriculture is both affected by climate change and contributed to climate change. Due to variation in rainfall, temperature, crops and cropping systems, the climate influences variably to different regions. Different crops also respond to climate change in different degrees. Since the tropical regions are more dependent on agriculture they are most affected by climate change. Rice, wheat, maize, sorghum, soybean and barley are the six major crops in the world grown in 40% cropped area, and contribute to 55% of non-meat calories and over 70% of animal feed (FAO, 2006). Hence success or failure of these crops will affect the major population of the world.

The negative impacts of climate change have been illustrated in Fig.4.8 by FAO (2006)

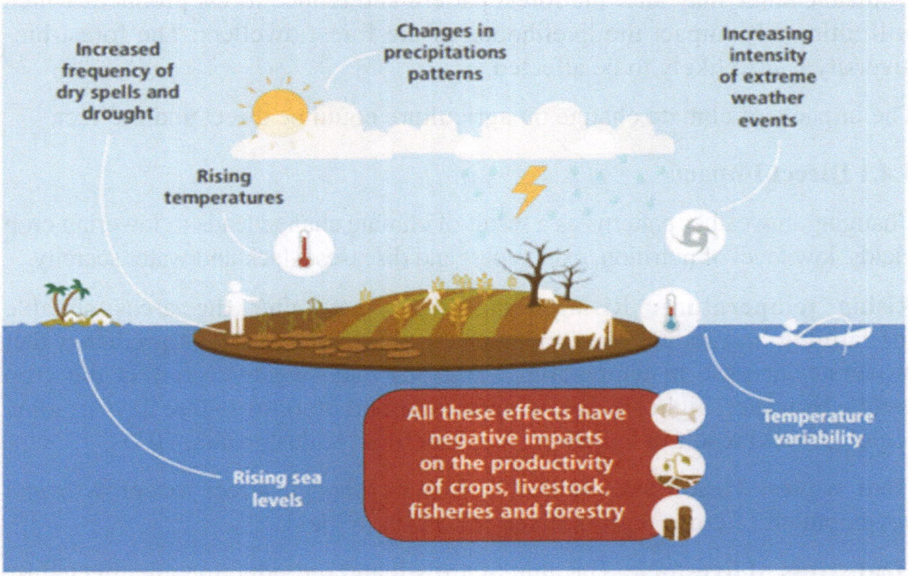

Fig. 4.8: Effects of climate change on agriculture

The IPCC has examined the published results from many different models and on the basis of the evidence has estimated that by 2100, the global temperature will rise by 1.1 to 6.4 °C and the sea level will rise between 18 and 59 cm. It is likely that hot extremes, heat waves and heavy precipitation will occur more frequently. Further, tropical cyclones will be more intense.

Studies by Indian Agricultural Research Institute (IARI) and others indicate greater expected loss in the Rabi crop and rise in temperature and rainfall have significant effects on the quality of fruits, vegetables, tea, coffee, aromatic and medicinal plants, and basmati rice. Both livestock production and fisheries production are also to be lowered. The reduction of south-west monsoon will affect the drinking water and irrigation water availability. With reduction in crop yields by direct and indirect impact of climate change there may be huge loss of revenue.

The loss in net revenue at the farm level is estimated to range between 9% and 25% for a temperature rise of 2 °C to 3.5 °C. Scientists also estimated that a 2°C rise in mean temperature and a 7% increase in mean precipitation would reduce net revenues by 12.3% for the country as a whole.

Change in climate may alter the distribution of disease causing parasites and other organisms which would threaten the human health.

Climate change may alter the forest pattern and reduce forest produces which will ultimately impact the livelihoods of the forest dwellers. The forest biodiversity is also likely to be affected.

The impacts of climate change on agriculture could be direct and indirect.

4.4.1 Direct Impacts

Changing in weather patterns as a result of climate change leads to lowering crop yields, low level of nutrition, excess pest and diseases attack and water scarcity.

Rising temperature: Rising temperature may alter the areas suitable for farming and reduce the crop yield in tropical areas. It is predicted that moderate increase in temperature (1–2 °C) may negatively affect the crop yields. However, higher winter temperatures and more frost-free days in some regions would result in longer growing seasons (Kulashrestha, 2011).

Heat waves: Heat waves linked to climate change affect the growth and development of crops with ultimate reduction in yield.

Heat stress of livestock: The animal growth and reproduction are affected by heat stress, which causes lower production of milk, egg and meat. Livestock of various species differ in their ability to cope with heat stress.

Changes in rainfall pattern: Changes in rainfall patterns cause droughts and floods which can destroy crops and eliminate food supply. This may disrupt agricultural activities and render the agriculture workers jobless. Instability in irrigation supply cannot support a good cropping. Due to less rainfall the rivers may be dry and the irrigation water would have to be transported from a longer distance involving huge cost. In drought situation the grazing lands are also affected which impacts livestock productivity. Recurrent drought situation compels the small farmers to leave farming and go on migration in search of job.

Carbon dioxide fertilisation: Carbon dioxide along with sunlight and water are important for photosynthesis. Elevated CO_2 increases crop yields and growth through an increase in photosynthetic rate, and it also decreases water loss as a result of stomatal closing (Hille, 2016). The CO_2 fertilization increases rate of photosynthesis while limiting leaf transpiration in plants. It is influenced by plant species, air, soil temperature and availability of water and nutrients (Hararuk, et al. 2018: Cartwright, 2013).

Reduced nutritional value of crops: Carbon dioxide levels in atmosphere affect level of protein and nutritional quality in crops. It is more pronounced in C_3 plants like rice, wheat, oat, etc. Food crops can show reduced level of protein, iron and zinc etc. due to changes in CO_2 level in atmosphere. Studies have shown that crops and wild plants exposed to elevated carbon dioxide levels at various latitudes have lower density of several minerals such as magnesium, iron, zinc, and potassium (Loladze, 2014).Increases in CO_2 lead to decreased concentrations of micronutrients in crop and non-crop plants with negative consequences for human nutrition including decreased B vitamins in rice (Zhu et al, 2018; Millus, 2018). A reduced level of magnesium, calcium, potassium, iron, zinc and other minerals in food crops can affect the human nutrition.

4.4.2 Indirect Impacts

Pest and disease attack: Global warming will alter pest, plant disease and weed distributions in crop plants which may reduce the crop yield in many crops. Warm and humid weather is always conducive for multiplication of insect pests and disease pathogens. Warm weather increases the metabolic rate and breeding cycle of insect populations. Climate change invariable increases the population and activities of specific pests like desert locust, fall armyworms, weeds, invasive species and plant pathogens. Huge crop loss is noticed by the attack of such insects and disease pathogens. It is found that cold temperatures at night and in the winter months would kill off insects, bacteria and fungi in many crops. High humidity, higher temperature and excess rainfall can favour the development of fungal diseases.

Food security, under-nutrition and food prices: It is invariably observed that climate change increases loss of food and production, which indirectly increases the food prices. Due to unavailability and loss of purchasing power and under-nutrition, food insecurity is seen. According to IPCC's Special Report on Climate Change and Land, food prices will rise by 80% by 2050 which will likely lead to food shortages. Rising food costs may pass on to the consumers and impacts the affordability of food, particularly in poor households. Food losses caused by fungi and bacteria due to climate change may also affect the issues of food safety.

Loss of agricultural land: Agricultural land adjacent to seas shores are inundated and washed away by the impact of rise in sea levels. Soil erosion, submergence of shorelines, salinity of the water table may increase due to rise in sea level. Low-lying areas near sea would experience more loss of land and crops like rice. Climate change may also increase the amount of arable land by reducing the amount of frozen land.

Reduced irrigation: Water runoff from glaciers that melt during the warmer summer months may be depleted due to climate change. Such depletion in runoff will affect the water availability in dams and reservoirs and this will cause low irrigation potential.

Erosion and soil fertility: Warmer atmosphere may impact hydrological cycle and cause excess rainfall. This may result more soil erosion and soil degradation. Unabated soil erosion may also cause loss of soil nutrients and soil carbon as a result of which soil fertility will be lost. Warming of soil due to climate change may cause increased soil microbe population size. Warmer conditions would favour growth of certain bacteria species in soil.

Change in flowering behaviour: Due to climate change the flowering behaviour of plants often changes. Sometimes the flowering times come earlier, and early blooms threaten the plants survival and reproduction. Early flowering increases the risk of frost damage in some plant species. Often there is mismatch between flowering and pollinators interaction. A long winter chill may also affect flowering in plants.

Impacts of surface level ozone on crops: Surface ozone layer caused due to methane emission acts as an air pollutant and reduces physiological functions and therefore the yield and quality of crops

4.4.3 Potential Effects of Climate Events on Agriculture

Projected impacts on crops

Increasing temperatures can damage the physical structure of soils, while increasing erosion and affecting soil fertility (FAO, 2013a). It is estimated that in the medium to long term climate-related changes will lead to negative effects on yields, effects that are likely to accelerate .These effects will be more severe in lower than at higher latitudes. The situation is particularly harmful for developing countries, whereas developed countries show a larger share of potential positive effects driven by climate change. Depending on the scenario used, researchers estimate that with no adoption, maize yields could decrease by up to 45 per cent, wheat by 50 per cent, rice by 30 per cent and soybeans by 60 per cent, compared with simulations that do not include climate change (FAO, 2016b). The projected impacts are indicated in Table 4.2.

Table 4.2: Projected Impacts on Crops

Cold periods becoming warmer and shorter; over most land areas, days and nights becoming hotter (almost certain).	Increased yields in colder environments; decreased yields in warmer environments; increased outbreaks of new insect pests and pathogens; potential impacts on crop production.
Heavy precipitation events increasing in frequency over most areas (very likely)	Damage to crops; soil erosion; inability to cultivate land owing to waterlogging of soil
Drought-affected area increases (likely)	Land degradation and soil erosion; lower yields from crop damage and failure; loss of arable land.
Land degradation and soil erosion; lower yields from crop damage and failure; loss of arable land.	Damage to crops.
Extremely high sea levels increase in incidence (excludes tsunamis) (likely	Salinization of irrigation water, estuaries and freshwater systems; loss of arable land.

Source: IPCC, 2006.

Impacts on livestock

Climate change would affect the productivity, health and feed of animals. Reduced availability of fodder may affect the productivity of milk and heat stress can increase the diseases in animals. Drought may affect the availability of fodder from pasture lands. While pasture productivity is increased by elevated CO_2 concentration, the quality may decrease due to high lignin concentration, low protein content and digestibility. Drought and waterlogging can create drinking water problem for animals.

Fisheries and aquaculture

Changes in temperature and seasons may affect the timing of reproduction and migration fishes. Capture fisheries and aquaculture systems will suffer from rising water temperatures and oxygen deficiencies, which may lead to change in productivity. In the oceans, sharp variations in climate trends will change the habitats of aquatic animals as well as the composition of fish species.

Forestry

Weather aberration limits the capacity of forests to produce goods and services as a result of which forest dwellers are most affected. Extreme climate events may alter the bio-diversity in forest eco-system.

References

Cartwright J (2013). "How does carbon fertilization affect crop yield?". environmentalresearchweb. Environmental Research Letters. Archived from the original on 27 June 2018. Retrieved 3 October 2016.

FAO, 2010, Climate-Smart Agriculture – Policies, Practices and Financing for Food Security, Adaptation and Mitigation, Presented at the Hague Conference on Agriculture, Food Security and Climate Change.

FAO. 2009. Low Greenhouse Gas Agriculture: Mitigation and Adaptation Potential of Sustainable Farming System.

FAO. 2012. Climate change adaptation and mitigation in agriculture. Course: climate change and food security. Rome

Hararuk O, Campbell EM, Antos JA, Parish R (December 2018). "Tree rings provide no evidence of a CO_2 fertilization effect in old-growth subalpine forests of western Canada". Global Change Biology.25 (4): 1222–1234. Bibcode:2019GCBio..25.1222H. doi:10.1111/gcb.14561. PMID 30588740.

Hille K (3 May 2016). "Rising Carbon Dioxide Levels Will Help and Hurt Crops". NASA. Retrieved 29 November 2018.

Impact of Climate Change on Agriculture and Food Security in India Int. Jr. of Agril., Env. and Biotech. Vol. 4, No. 2 : June 2011 : 129-137

IPCC (Intergovernmental Panel on Climatic Change) 2006. The Economics of Climate Change: Stern Review. The Summary of conclusions. Survey of the Environment 2007, The Hindu, pp141-145.

IPCC (Intergovernmental Panel on Climatic Change) 2006. The Economics of Climate Change: Stern Review. The Summary of conclusions. Survey of the Environment 2007, The Hindu, pp141-145.

IPCC (Intergovernmental Panel on Climatic Change) 2007. Climate Change: The Physical Science Basis. Extracts from the IV Assessment Report. Survey of the Environment 2007, The Hindu, pp147-155.

IPCC (Intergovernmental Panel on Climatic Change) 2007. Climate Change: The Physical Science Basis. Extracts from the IV Assessment Report. Survey of the Environment 2007, The Hindu, pp147-155.

IPCC. 2001. Third Assessment Report of the Intergovernmental Panel on Climate Change. Cambridge University Press, Cambridge

Kim Chang-gil. 2009. "Strategies for Implementing Green Growth in Agricultural Sector." in Proceedings in Green Korea 2009 - Green Growth and Cooperation. National Research Council for Economics, Humanities and Social Science.

Kulshreshtha SN (March 2011). «Climate Change, Prairie Agriculture and Prairie Economy: The new normal». Canadian Journal of Agricultural Economics. 59 (1): 19–44. doi:10.1111/j.1744-7976.2010.01211.x

Loladze I (May 2014). "Hidden shift of the ionome of plants exposed to elevated CO_2 depletes minerals at the base of human nutrition". eLife. 3 (9): e02245. doi:10.7554/eLife.02245. PMC 4034684. PMID 24867639.

Milius S (23 May 2018). "As CO2 increases, rice loses B vitamins and other nutrients". Sciencenews.org. Retrieved 2 July 2018

OECD. 2006b. The Impact of Climate Change on Agriculture and Options for Adaptation. COM/AGR/CA/ENV/EPOC(2006)2

Paul Erhlich on climate change in 1968. Backseat driving. Retrieved November 17, 2005. pages 51–52 of The Population Bomb, 1968, available from Schmidt, Brian (July 1, 2005).

Rasool, S.I.; Schneider, S.H. (July 9, 1971). "Atmospheric Carbon Dioxide and Aerosols: Effects of Large Increases on Global Climate". Science. 173 (3992): 138–41

World Bank. 2008. World Bank's Approach to Climate Change in South Asia: An Overview, Bank Information Center, online available at: www.bicusa.org

Zhu C, Kobayashi K, Loladze I, Zhu J, Jiang Q, Xu X, et al. (May 2018). "Carbon dioxide (CO_2) levels this century will alter the protein, micronutrients, and vitamin content of rice grains with potential health consequences for the poorest rice-dependent countries". Science Advances. 4(5): eaaq1012. Bibcode:2018SciA....4.1012Z. doi:10.1126/sciadv.aaq1012. PMC 5966189. PMID 29806023.

5

Challenges of Climate Change to Sustainable Food Security

5.1 Agriculture and Food Security

Food security refers to a situation when people gain physical and access to enough food to meet the dietary requirements for a healthy life for all times. A household can have food security when all the members lead a hunger free life. It is estimated that nearly 800 million people in the world go to bed hungry every night. Food security exists when all people, at all times, have physical and economic access to sufficient, safe and nutritious food that meets their dietary needs and food preferences for an active and healthy life (World Food Summit, 1996). This definition gives rise to four dimensions of food security: availability of food, accessibility (economically and physically), utilization (the way it is used and assimilated by the human body) and stability of these three dimensions.

(i) **Availability**: The availability of sufficient quantities of food of appropriate quality is achieved through domestic production, distribution and through exchanges. When domestic production falls short of requirement the country manages the situation by import of food aid. The food available needs proper distribution so as to reach all needy households.

(ii) **Access**: Physical, economic and environmental accesses are very important for ensuring food security. The individuals should have adequate resources for buying food and the food they require, which is called their purchasing power. They should also exercise their entitlements to have physical access to food. Food subsidies and food aids are often provided to improve the economic access to food.

(iii) **Utilization or food use**: With a view to meeting the physiological needs, the individuals should have access to adequate diet, clean water, sanitation and healthcare. Nutrient content of food, food safety and human health are required for proper utilisation of food.

Box 1: Food security

"Food security exists when all people, at all times, have physical and economic access to sufficient, safe and nutritious food that meets their dietary needs and food preferences for an active and healthy life" (World Food Summit, 1996). This widely accepted definition points to the following four dimensions of food security:

Availability: The availability of sufficient quantities of food of appropriate quality, supplied through domestic production or imports (including food aid).

Access: Access by individuals to adequate resources (entitlements) for acquiring appropriate foods for a nutritious diet. Entitlements are defined as the set of all commodity bundles over which a person can establish command given the legal, political, economic and social arrangements of the community in which they live (including traditional rights such as access to common resources).

Utilization: Utilization of food through adequate diet, clean water, sanitation and healthcare to reach a state of nutritional well-being where all physiological needs are met.

Stability: To be food secure, a population, household or individual must have access to adequate food at all times. They should not risk losing access to food as a consequence of sudden shocks (e.g. an economic or climatic crisis) or cyclical events (e.g. seasonal food insecurity). The concept of stability therefore refers to the availability, access and utilization dimensions of food security.

(iv) Stability: Stability relates to all time availability, access and utilization of food. To be food secure, a population, household or individual must have access to adequate food at all times. At any time they should not face the consequence of sudden shocks or cyclic events like seasonal food insecurity.

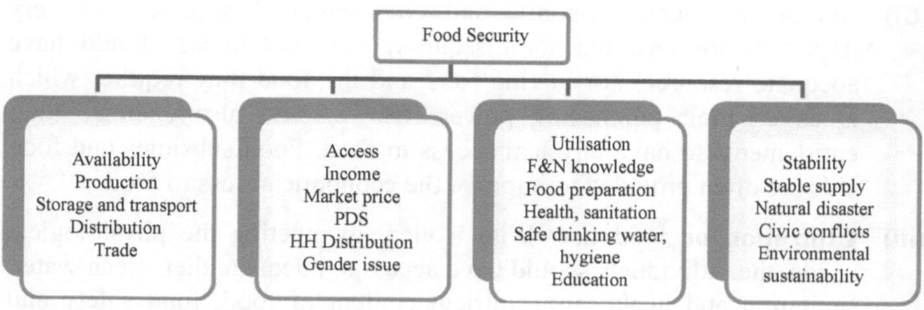

Fig. 5.1: Dimensions of food security

Agriculture has a major role in achieving food security of a nation. Insufficient food production and access, climate change, natural disasters, conflicts, displacement, and population explosion are among the many challenges faced by the families and communities to secure food security. A situation of food insecurity is also a cause of migration. Usually the families facing food insecurity rely on their own food production and income from farming. But weather-related problems triggered by climate change impact their food security. Agriculture and allied sectors like animal husbandry, fisheries and forestry are sensitive to climate change. The overall impacts of climate change on agriculture and allied sectors are negative. Invariably agriculture faces water scarcity, natural hazards like drought, flood, cyclone etc. due to the impact of climate change as a result of which food security is threatened. Since agriculture is major source of livelihood of rural families, climate change also affects the livelihoods of these people.

5.2 Impact of Climate Change on Food Security

The impacts of climate change on food security is summarised below.

Dimension	Impact of climate change
Availability	Climate change may cause such weather events which are not conducive for agricultural production. Declines in the availability of wild foods, and limits on small-scale horticultural production due to scarcity of water or labour resulting from climate change could affect nutritional status adversely. Climate change will cause new patterns of pests and diseases to emerge, affecting plants, animals and humans, and posing new risks for food security, food safety and human health.
Access	Climate change may impact income-earning opportunities which would affect the ability to buy food, and a change in climate or climate extremes may affect the availability of certain food products, which may influence their price. High prices may make certain foods unaffordable and can have an impact on individuals' nutrition and health.
Utilisation	Climate change may affect both the physical and the economic availability of certain preferred food items, which might make it impossible to meet some preferences. Changes in availability and relative prices for major food items may result in people either changing their food basket, or spending a greater percentage of their income on food when prices of preferred food items increase. Increased incidence of water-borne diseases in flood-prone areas, changes in vectors for climate-responsive pests and diseases, and emergence of new diseases could affect both the food chain and people's physiological capacity to obtain necessary nutrients from the foods consumed.

Stability	Droughts and floods are a particular threat to food stability and could bring about both chronic and transitory food insecurity. Global food markets may exhibit greater price volatility, jeopardizing the stability of returns to farmers and the access to purchased food of both farming and non-farming poor people. Higher temperatures and humidity associated with climate change may require increased expenditure to preserve stored grain, which will limit countries' ability to maintain reserves of sufficient size to respond adequately to large-scale natural or human-incurred disasters.

With the current population growth consideration, it is predicted that the population of world may reach nine to ten billions by 2050. It may present a tough challenge to produce enough food for such predicted population. The possible options could be closing the yield gap, raising the yield ceiling of crops through better technologies, reducing food wastes and control of population. Increasing food production would require reduction of climate impacts on food production and increasing resilience of agriculture to climate change in future. The present agriculture is constrained with anthropogenic pressure, shrinking resources like land and water and impact of climate variability.

All possible steps are being taken by all nations to ensure food security with due regard to all four dimensions. Land use for production of food and food security are intimately linked in the context of climate change. Food production is only one challenge leading to food availability, but other aspects like access and utilisation are governed by other factors.

Food production from agriculture is extremely dependent on temperature and rainfall and therefore vulnerable to climate change (Nelson et al, 2009). Therefore, the climate change impacts on agriculture could be mostly negative and it may always threaten the food security. The populations in developing countries, which are already food insecure and vulnerable to climate change, are likely to be the worst affected. Climate change will not only reduce the crop yields affecting food availability, but also it may increases food prices leading to low consumption. In fact, climate change may challenge all four dimensions of food security. Physical, biological and biophysical impacts of climate change may influence agro-ecosystems and food production. It may lead to quantity, quality and price effects, with impacts on the income of farm households and on purchasing power of non-farm households. As the income and employment of poor households will be affected by climate change they may reduce the consumption expenditure along with reduced expenditure on health care and education. Stability of food supply will be impacted by changes in seasonality, increased variance of ecosystem productivity, increased supply risks and reduced supply predictability – issues that may also have large impacts on supply chain costs and retail prices (FAO, 2008). Hence all four dimensions of food security (availability, access, utilisation and stability) and nutrition are likely to be impacted by these effects (Fig.5.2).

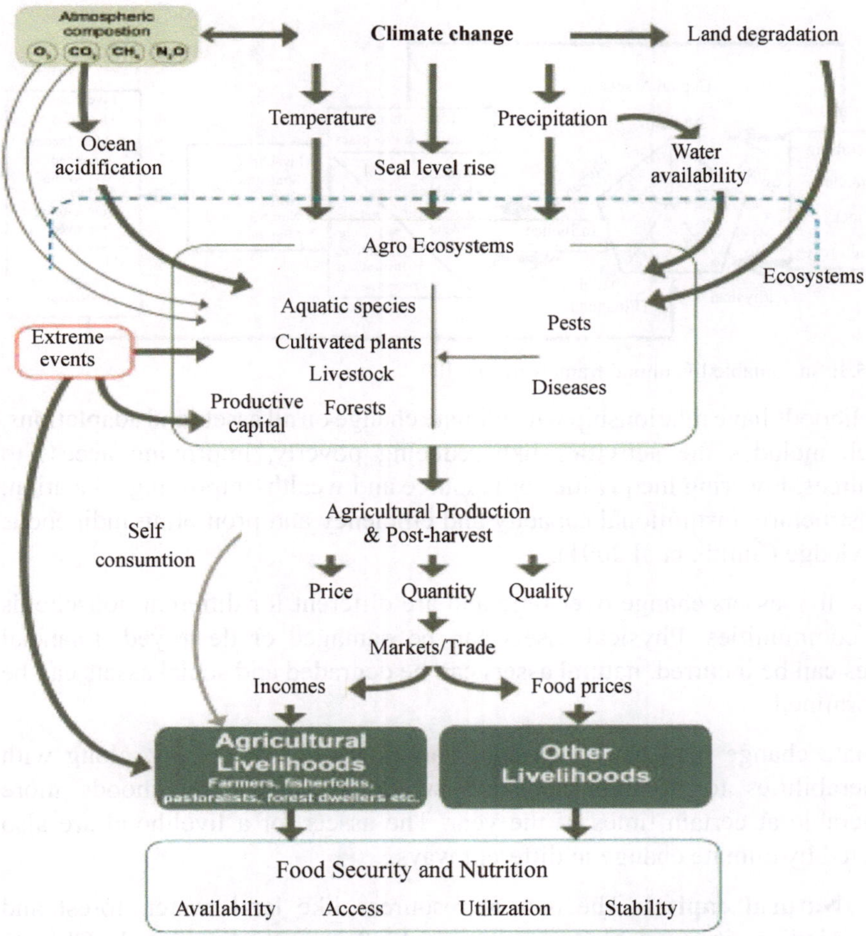

Fig. 5.2: Effects of climate change impacts on food security and nutrition.

5.3 Climate Change and Livelihoods

The livelihood of a person refers to "means of securing the basic necessities of life." Livelihood comprises the capabilities, assets (including both material and social resources), and activities required for a means of living (DFID, 2002). The five assets essential for person's livelihood are physical, financial, natural, social and human.

A livelihood is sustainable when it can cope with and recover from stresses and shocks and maintain or enhance its capabilities and assets both now and in the future, while not undermining the natural resource base" (Carney 1998, page 4). A sustainable livelihood framework of DFID explains these five assets required for livelihood of a person (Fig.5.3).

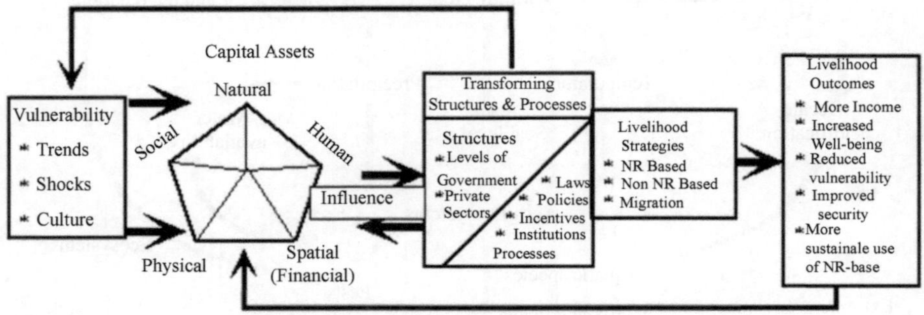

Fig. 5.3: Sustainable livelihood framework (DFID)

Livelihoods have relationship with climate change on all assets and adaptations, which includes the activities like reducing poverty, improving access to resources, lowering inequalities of resource and wealth, improving education, infrastructure, institutional capacity and efficiency and promoting indigenous knowledge (Smith, et al 2001).

These five assets change over time and are different for different households and communities. Physical assets can be damaged or destroyed, financial losses can be incurred, natural assets can be degraded and social assets can be undermined.

Climate change may cause seasonal variations in food supply, along with vulnerabilities to flooding and fire, which can make livelihoods more vulnerable at certain times of the year. The assets for a livelihood are also affected by climate change in different ways.

(i) **Natural capital**: The natural resources like land, water, forest and biodiversity available to an individual influence the livelihoods. Climate change leads to increase of arid lands, causes land degradation and erosion, and water stress while prolonged drought or waterlogging alter the physical, chemical and biological properties of soil. The growth in forest ecosystem is reduced due to climate change. All these changes put a challenge to the livelihood security of people.

(ii) **Social capital**: Natural events like heavy rains, heat waves and cold waves caused by climate change have negative effects on products and services of social assets like groups and networks. Climate change often leads a situation of forced migration of marginal and poor people.

(iii) **Human capital**: In individual households, human capital like skills, knowledge, ability to labour and good health in terms of quantity and quality are also affected by weather events caused by climate change. The adverse effects of climate change reflected in food insecurity, undernourishment and chronic hunger, due to failed crops or price rise of food affect the human capital like skill, knowledge, ability to labour and good health of individuals. Climate change is believed to have different effects on men and women; in some cases, it leaves women more vulnerable and disadvantaged than men (Mainlay & Tan, 2012).

(iv) **Physical capital**: The basic infrastructure for transport, buildings, water management, energy, and communications and productive capital (tools, machines, etc.) which enables people to pursue the livelihoods. Physical assets are provided by government or private sectors to support livelihoods of people. Such assets are affected by disasters like floods, landslide and other physical events. Losses of physical assets have multiplier effects on economy and livelihoods of people dependent on them.

(v) **Financial capital**: The financial resources like savings, supplies of credit, regular remittances and pensions, social security payments or insurance are essential for different livelihood options of people. Natural disasters and extreme weather events reduces the labour outputs and working hours. Climate change affects agricultural profitability due to application of excess labour and inputs.

Exploitation of common property resources (CPR) for livestock, fishing, gathering fuel wood and many other things is required by poor people to ensure their livelihood security. Climate induced changes to resource flows (whether temporary, reflecting variability or structural, reflecting change) can affect these resources also.

5.4 Achieving Food Security through Adaptation and Mitigation of Climate Change

Adaptation and mitigation of climate change are required to address the challenges of climate change to food and livelihood security of people. Adaptation refers to some measures of adjustments to decrease the vulnerability of production systems to climate changes. Mitigation is the effort to reduce the adverse impact of climate change. Many technologies and practices have been evolved for enhancing the ability of crop production system to adapt and mitigate the adverse effects of climate change. Adaptation strategies focus on disaster risk management and risk transfer mechanisms like diversified livelihood systems, integrated crop-livestock and aquaculture-agriculture systems and crop insurances. Rao et al. (1989) found that in India, technological innovations in conjunction with input management and adoption of management practices result in stability of sorghum and millet cultivated during the rainy season. Virmani (1989) found that improved cropping systems can cope best with rainfall variability when applied concurrently with an efficient set of soil and water management techniques.

5.4.1 Adapting to Climate Change

Adaptation to climate change includes good practices to strengthen the resilience to climate change and management of risks by required information and crop insurance coverage. It also seeks new institutional and technological solutions against adverse impacts of climate change.

Components of climate change adaptation

It is essential to adopt an integrated approach to manage climate risks at various levels. The activities to promote adaptation and reduce vulnerability require following strategy elements in bottom-up manner as suggested by FAO, 2006 (Fig.5.4)

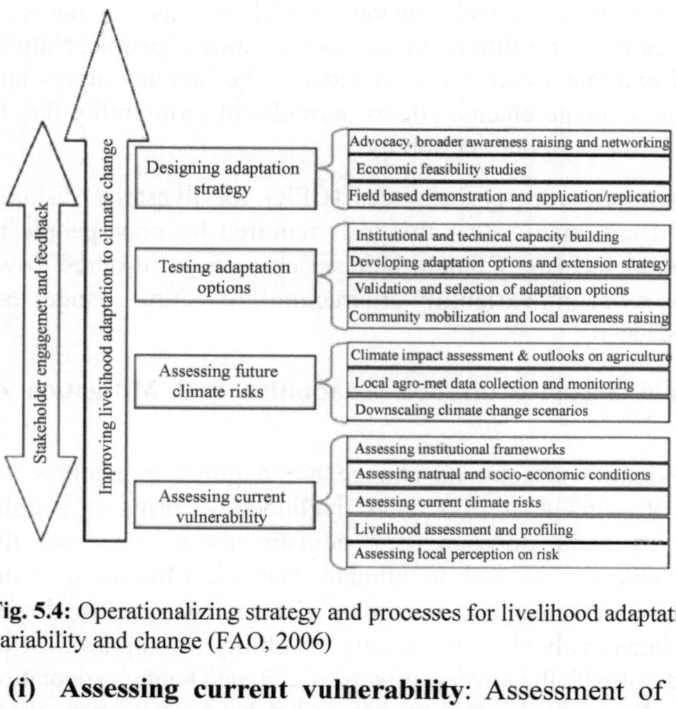

Fig. 5.4: Operationalizing strategy and processes for livelihood adaptation to climate variability and change (FAO, 2006)

(i) **Assessing current vulnerability**: Assessment of natural and socio-economic conditions, current climate risks, local perception about climate risks and impacts, and documentation of livelihood profiles along with institutional framework would be required for the pilot sites. These information will be useful to understand the vulnerability to climate risk.

(ii) **Assessing future climate risks:** Assessment of impact of climate change, agriculture scenario and local agro-meteorological data, which would form the basis for future climate risks in the area.

(iii) **Testing adaptation options:** Identification and selection of viable options for adaptation are to be made followed by validation and preparation of farmer-friendly practices along with extension strategy. Community mobilisation will have to be made for raising awareness.

(iv) **Design adaptation strategy**: Adaptation strategy for dissemination and replication of successful options will be prepared by awareness raising, economic feasibility studies and filed-based demonstration and applications.

Development of an adaptation options menu

Adaptation options for managing climate risks are to be identified for synthesizing adaptation practices. The four steps to be followed for such selection of adaptation processes are displayed in Fig.5.5 (FAO, 2006). These steps are within the overall adaptation policy of a country.

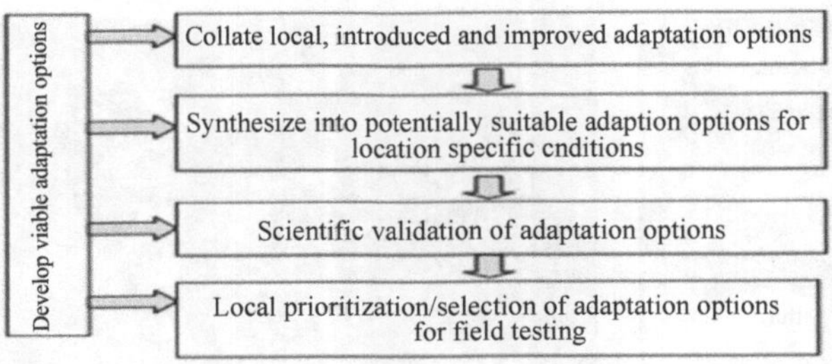

Fig. 5.5: Sequential tasks in designing viable adaptation options (FAO, 2006)

5.4.2 Mitigation of climate change

Food and livelihood security can be protected by mitigating climate change. Conversion of forest land to pasture or cropland, and intensive crop and livestock production practices have contributed much greenhouse gas emissions. Therefore appropriate practices to reducing emissions of CO_2 and other GHGs have to be adopted.

(i) Reducing carbon dioxide emissions from agricultural and forestry: Conversion of land from forested land to cultivated or grazing land is the primary source of CO_2 emission. It can be reduced by sustainable farming practices, strengthening ecosystem resilience and more efficient energy use. The forest fires need to be controlled and stubble burning should be discouraged. UNFCCC and the Kyoto Protocol recognize the potential role of forests in providing a variety of adaptive ecosystem services in addition to mitigating climate change through carbon sequestration. Afforestation, agro-forestry and integrated farming systems with tree component are sound practices to reduce CO_2 emissions.

(ii) Reducing emissions of methane and nitrous oxide from agriculture: The enteric fermentation of digested fibres by large ruminants such as cattle, sheep and goats is an important source of release of methane from livestock farming. Rice cultivation under flooded condition is another source of methane emission. Nitrous oxide is emitted from manure, cultivated soils that have been fertilized with organic matter or inorganic compounds containing nitrogen, and nitrogen fixing legumes. Grazing and feed management in livestock farming, manure management, and changing the cropping pattern and adopting integrated farming system are viable options to reduce the emission of these greenhouse gases.

The measures required for reduction of GHG from forestry, livestock farming and agriculture are displayed in Fig.5.6, 5.7.5.8.

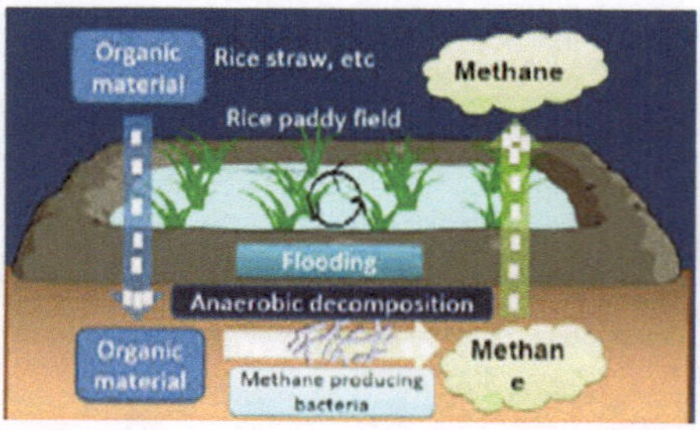

Fig. 5.6: Reducing methane from rice field

Fig. 5.7: Grazing management in ruminants

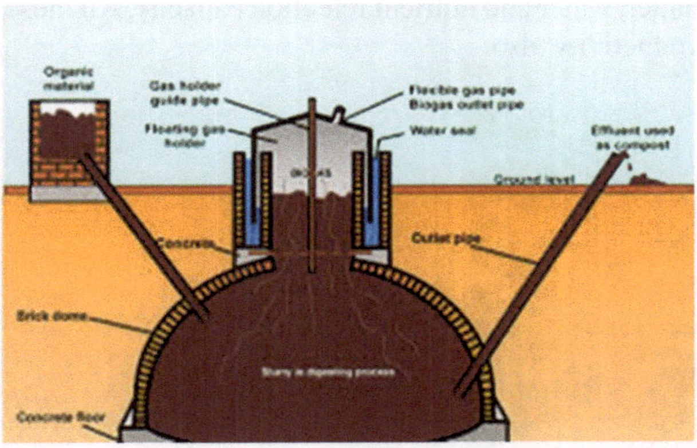

Fig. 5.8: Preparation of biogas from cow dung to reduce methane emission

(iii) Conservation agriculture

Conservation agriculture involving zero tillage/minimum tillage, mulching, cover cropping etc. is an option for adaptation as well as for mitigation of climate change. In conservation agriculture mechanical soil tillage is reduced to an absolute minimum, and external inputs such as agrochemicals and nutrients of mineral or organic origin are applied at optimum levels .Under conservation agriculture, no soil moisture is lost through tillage and seedbed preparation. Conservation agriculture can significantly save energy compared with conventional tillage. The crops grown under conservation agriculture are found to be drought tolerant. Yield fluctuations under conservation agriculture are generally much less severe than under comparable conventional agriculture (Tebrügge and Bohmsen, 1998; Derpsch, 2005).

(iv) Biodiversity conservation

Genetically diverse populations and species-rich ecosystems can adapt to climate change better. FAO promotes the use of indigenous and locally adapted plant and animal diversity, and the selection and multiplication of crop varieties and autochthonous races that are adapted or resistant to adverse conditions. Breeding plants and animals for tolerance to drought, heat stress, salinity and flooding will also become increasingly important.

(v) Carbon sequestration

Carbon sequestration involves increasing the carbon storage in terrestrial systems, above or below ground. Planting trees to serve as carbon sinks, changing soil management practices can sequester carbon and reduce emissions. Management practices for carbon sequestration have additional benefits, including increased root biomass, soil organic matter, water and nutrient retention capacity. All these benefits increase the land productivity also.

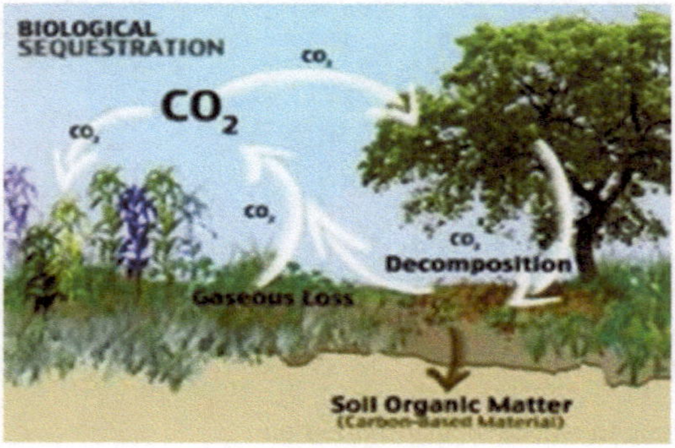

Fig. 5.9: Biological sequestration

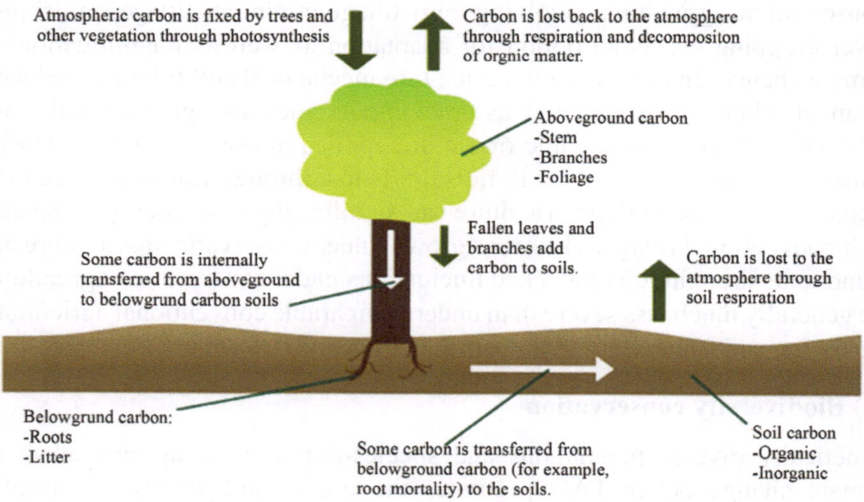

Fig. 5.10: Carbon sequestration in soil ecosystem

5.5 Climatic Solution Farming

The objective of climate solution farming is to find out ways that can significantly lower greenhouse gas emissions, store additional carbon from the atmosphere in the soil, and reduce vulnerability to the effects of climate change. Instead of using chemical fertilisers increased use of animal manures, compost, green manure crops that are incorporated into the soil to provide nutrients and organic matter along with crop rotation are to be adopted.

Although these practices may not fully replace synthetic fertilizer use, especially with nitrogen-dependent crops, they can significantly reduce the quantity of fertilisers needed. Another solution is through perennial cropping systems like agro-forestry, tree planting and fruit cultivation. These crops have multiple benefits like significantly reducing the need for tillage; deeper root systems that both protect and build soil; better drought tolerance; and lower fertilizer and pesticide requirements. Putting more carbon in the soil can help reduce the damaging levels of these gases in the atmosphere and mitigate some of the harmful impacts on agricultural production of our already changing climate. This can be done by no-till farming, crop rotation and cover cropping etc.

References

Carney, D., ed. Sustainable Rural Livelihoods. London: DFID, 1998.

Derpsch, R. 2005. The extent of CA adoption worldwide: Implications and impacts. Paper presented at III World Congress on Conservation Agriculture, Nairobi.

EuropaWorld. 2001. Conservation agriculture can benefit the planet, says FAO. Available at: FAO (2008) The State of Food Insecurity in the World 2008 .

FAO. 2004a. Carbon sequestration in dryland soils. World Soil Resources Reports No. 102. Rome. Available at: ftp://ftp.fao.org/agl/agll/docs/wsrr102.pdf.

Mailnay, J & Tan, S.2012. Mainstreaming gender and climate change in Nepal. London, UK: IIED Climate Change Working Paper No.2

Nelson GC, Rosegrant MW, Koo J et al. (2009) Climate Change. Impact on Agriculture and Costs of Adaptation, Climate change and sustainable food production . Proceedings of the Nutrition Society p. 19. Washington, DC: International Food Policy Research Institute, pp. 19.

Rao, N. G., P., Rao, G. R., K. and Acharya, H. S. (1989). Yield Stability of Sorghum and Millet Across Climates. In: Climate and Food Security. International Symposium on Climate Variability and Food Security in Developing Countries 5-9 February 1987. Manila, Philippines: International Rice Research Institute and Washington, D.C.: American Association for the Advancement of Science

Tebrügge, F. & Bohmsen, A. 1998. Farmers' and experts' opinion on no-tillage: Results of a Europewide survey. In Proceedings of the Conference of Agricultural Engineering 1998. Düsseldorf, Germany, Association of German Engineers-Society of Agricultural Engineering (VDI-MEG).

Virmani, S. M. (1989). Cropping Systems Strategies for Coping with Climatic Fluctuations. In: Climate and Food Security. International Symposium on Climate Variability and Food Security in Developing Countries 5-9 February 1987. Manila, Philippines: International Rice Research Institute and Washington, D. C.: American Association for the Advancement of Science

World Food Summit. 1996. Rome Declaration on World Food Security. Rome, FAO www.europaworld.org/issue/conservation51001.htm.

6

Climate Resilient Agriculture

Climatic factors like increase in temperature, carbon dioxide (CO_2) concentration, moisture, and other factors directly or indirectly affect agriculture, which have been elaborately dealt in previous chapters. Climate change causes weather variability that may result in low production, hike in price of agricultural commodities including food grains, vegetables, fruits, meat, egg and milk. It is now realised that climate resilient agriculture (CRA) will be imperative to reduce the vulnerability of agriculture to the climate change.

6.1 Climate Resilient Agriculture (CRA)

CRA means the incorporation of adaptation, mitigation and other practices in agriculture to address the adverse effects of climate change. It includes judicious and better management of natural resources through adoption of new technologies and practices. Climate-resilient agriculture practices can alter the current problems of climate change and sustain agricultural production from the local to the global level, especially in a sustainable manner.

Climate resilient agriculture is an integrated approach that aims at three objectives: (i) sustainably increasing agricultural productivity, to support equitable increases in farm incomes; (ii) adapting and building resilience of agricultural and food security systems; and (iii) reducing emissions of greenhouse gasses from agriculture (including crops, livestock and fisheries) to the possible extent.

6.2 Climate Resilient Practices

Various practices to be adopted for climate resilient agriculture can be grouped under agriculture sub-sectors following the guidelines of FAO, 2021.

6.2.1 Crops and Cropping Systems

The crops grown under different situations are usually affected by extensive heat, strong wind, cold waves, hailstorms, drought, flood& waterlogging, land degradation, greenhouse gas emission, salinization and pest attack. Various practices against each of such disruptions are outlined below with possible benefits.

Extreme heat: Use of heat tolerant crops (e.g. Millets) and varieties with short maturity period and adjustment of sowing time to avoid heat stress period can reduce the effects of heat and increase crop yields.

Drought: The resilient practices include adoption of agro-forestry and alternate cropping systems, selection of drought tolerant crops and varieties, adjusting planting seasons to minimise the stress exposure, construction of terraces, mini soil and water conservation measures like half-moon terrace, staggered trenches, farm ponds, broad bed and furrow system (BBF), provision of life saving system from water harvesting structures, rain water harvesting, timely weeding, cover cropping, mixed and intercropping etc. to reduce the impact of drought.

Cold, frost and hail: Frost protection can be made by planting tree rows and frost protection irrigation. Soil mulching increases surface temperature and canopy cover can provide protection against frost damages by reducing net radiation losses. Early warning and adjusting time of planting may reduce the impact of cold and frost.

Flood and waterlogging: Provision of drainage facilities, raised bed system of cultivation, field dredging, agronomic measures like sub-soiling, application of organic manures, and provision of bio-drainage can reduce the impact of flooding and waterlogging. Crop planning may also be made as per field situations and period of water logging. Use of flood tolerant crops and varieties can reduce the adverse impact of water logging and flooding.

Land degradation: Crop rotation, cover cropping, cultivation of legume crops, contour farming, strip cropping, mulching, minimum tillage, application of organic matter, early sowing of crops to allow a vigorous root system and adoption of soil conservation measures can protect the land degradation and make the cropping resilient to climate change.

Reducing emissions of GHG: Adoption of wetting drying technology in rice cultivation, direct seeding, system of rice intensification (SRI) , manure management to reduce the methane emission, carbon sequestration in soil and in tree farming, carbon fixation in C_3, C_4, CAM plants, precision agriculture, site specific nutrient management of plant nutrients by using greenSeeker and leaf colour chart (LCC), integrated nutrient management, and increasing energy use efficiency etc. are proven technologies to reduce the emissions of GHG from agriculture.

Salinization: Promotion of halophyte crops, use of saline tolerant crops and varieties, sub-soil drainage, soil amendment by application of gypsum, measures against vertical movement of saline water, construction of barriers to restrict saline inundation from sea water, seed bed management and field grading and management of irrigation system are resilient practices against climate change.

Management of crop pests and diseases: Integrated pest management involving host plant resistance, cultural control, physical and mechanical control and biological pest control measures are required to limit the pests and diseases within the economic injury level. The crop management practices should be so designed that it would encourage the activities of natural enemies to control the harmful organisms in the field condition. Biological pest control using predators, parasites, nuclear polyhedrosis virus (NPV), botanical pesticides like neem and karanj , various traps like pheromone traps, yellow traps, sticky traps, and regular pest surveillance in crop fields are very useful and effective against pest and disease attack.

6.2.2 Livestock System

Livestock production is affected by climate change as a result of which the production of milk, egg and meat is reduced and the cost of such products goes on increasing. The climate resilient practices to reduce the effects of climate change on livestock system are described below.

Drought and heat: In order to save the livestock from drought and heat, the measures like provision of drinking water, reduction of overgrazing, introduction of breeds tolerant to heat stress, provision of shelter, integration of livestock with agriculture and tree farming, provision of shelter belts, bio-diversity conservation and reduction of animal mortality by reducing the exposure to heat stress etc. can be adopted.

Land degradation: The resilient practices like rotational grazing, fodder management, soil management, reduction of herd size, introduction of silvi-pastoral farming system, livestock based integrated farming system can be adopted to reduce land degradation for livestock farming.

Emission of greenhouse gases: Practices to reduce enteric fermentation in ruminants by improving nutrition, manure management and bio-gas production, appropriate plant breeding to improve forage qualities, diet supplementation, integrated farming system etc. are to be adopted.

Animal diseases: Preventive measures against animal diseases, control of flies and mosquitoes in animal sheds, production of quality fodder and use of various traps are necessary to be adopted as practices for climate resilience.

6.2.3 Forest System

Degradation of forest land: A stock-taking exercise has to be made before forest planning. Forest inventory is required to assess the forest growing conditions, forest stock resources, forest hazards before forest planning. Sustainable management of forest ecosystem, biodiversity conservation, control of grazing, controlled timber harvesting, control of forest fire, water conservation, conservation of top soil, afforestation etc. are viable options.

Emission of greenhouse gases: Reforestation and afforestation for carbon sequestration, control of forest fire by stone fencing, restricting shifting cultivation and trash burning, nutrient management, increasing soil moisture etc. are required to limit emission of greenhouse gases.

Drought: Sustainable forest management, enhancing biodiversity to enhance resilience to dry conditions, mixed plantation, avoiding clear-cut logging and preserving the forest biome, soil and water management interventions may reduce the impact of drought.

Pest and disease: Pest monitoring, cultural management like pruning, planting of resilient varieties, removal of pest from the habitat by introducing mechanical traps, screens, fences and nets , use of beneficial predators, parasites and pathogens to control pests, and use of natural pesticides are required to control pests and diseases.

6.2.4 Fisheries and Aquaculture

Limited catching of fish from sea, restoration of mangrove plantation, reducing coastal soil erosion, reducing ocean acidification, disaster management with early warning, construction of seawalls and flood reservoirs, or buffer zones via afforestation or mangrove reforestation, breeding management, introduction of locally available fish species in aquaculture, improved pond management, feed management and increasing resource use efficiency are various options in fisheries management against climate change.

6.3 Climate resilient agriculture through adaptation and mitigation strategies

National Innovations in Climate Resilient Agriculture (NICRA) was launched by Indian Council of Agricultural Research (ICAR) in 2011. The basic objectives of the flagship programme were:

- To undertake strategic and applied research on climate change adaptation and mitigation;

- To validate, demonstrate and assess the impact of climate resilient technologies on farmers' fields; and
- To strengthen the capacity of scientists, farmers and other stakeholders on climate resilient agriculture.

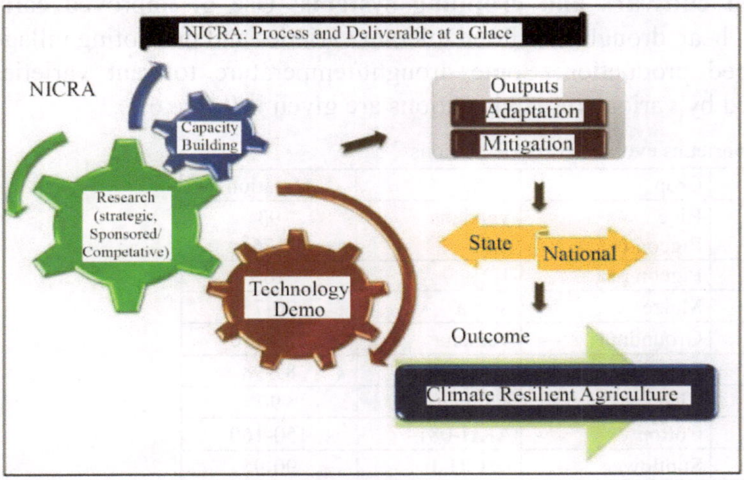

Fig. 6.1: Process and Deliverables of NICRA

The project is implemented in one representative Gram Panchayat in each of the selected 100 districts in the country basing on major climate vulnerability by the Krishi Vigyan Kendras (KVK) under the overall guidance of Central Research Institute on Dryland Agriculture (CRIDA), Hyderabad (Fig. 6.2).

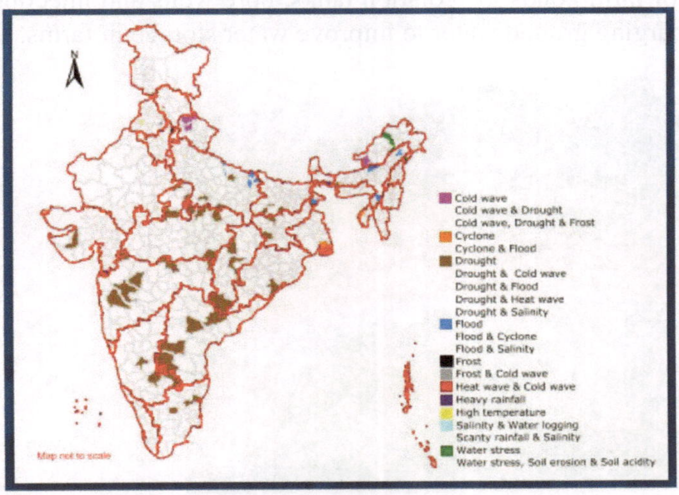

Fig. 6.2: Selected GPs under NICRA project

The following village level interventions were made in the project area.

(i) **Building resilience in soil**: Building soil carbon, control of soil erosion, enhancing water holding capacity of soil, and balanced nutrition to maintain soil health.

(ii) **Adapted cultivars and cropping systems**: Use of improved early duration heat, drought and flood tolerant varieties and promoting village level seed production. Some drought/temperature tolerant varieties identified by various research stations are given in Table 6.1:

Table 6.1: Crop varieties evolved at ICAR Stations

Location	Crop	Variety	Duration
Varanasi	Rice Pigeon pea	Vandana T.21	93 176
Phulbani	Pigeon pea	T.21	160
Arjia	Maize	Surya	70-75
Anatapur	Groundnut	Vemana	105-110
Indore	Soybean	JS-90-41	87-98
Rewa	Rice	Kalinga-3	90
Akola	Cotton	AKH-081	150-160
Bijapur	Sunflower	KBSH-1	90-95
Solapur	Sorghum	Mauli	105-110
Hisar	Pearl millet	HHB-67	60-62
Bangalore	Finger millet	GPU-26	90-105

Source; CRIDA, Hyderabad

(iii) **Rainwater harvesting and recycling**: Restoration of old water bodies, construction of farm ponds, percolation tanks, bore wells and injection wells for recharging ground water to improve water storage in farms.

Fig. 6.3: A model farm pond for rainwater harvesting

(iv) **Water saving technologies:** Introduction of water saving technologies like zero tillage, direct seeded rice and other resource conservation technologies to save water and reduce GHG.

(v) **Farm machinery (custom hiring) centres**: Improving access to farm mechanisation for timely land preparation through community managed custom hiring centres.

(vi) **Crop contingency plan:** ICAR/CRIDA have prepared contingency plan for 400 districts of the country to cope with climate change.

(vii) **Livestock and fisheries interventions**: Use of community land for fodder production, storage of fodder at village level, feed supplements to reduce nutrient deficiency, preventive vaccination, livestock shelters to reduce heat/drought shock in animals and fish management to improve productivity.

Fig. 6.4: Livestock management at village level

(viii) **Weather based agro-advisories:** Automated weather stations at KVKs, mini weather recording stations at project villages to provide weather-based crop advisories to farmers.

(ix) **Institutional interventions:** Initiation of seed bank, fodder bank, custom hiring centre, collective marketing, and weather index based crop insurance to improve climate literacy at village level.

(x) **Village Climate Risk Management Committee (VCRMC):** Promotion of farmers' participation through a VCRMC constituted taking all types of farmers including women and landless with the approval of Gram Sabha for taking all decisions on climate resilient agriculture.

Fig. 6.5: VCRMC meeting

6.4 Key Technologies/Strategies for Climate Resilient Agriculture

The climate resilient technologies could be grouped under following categories for convenience of planning and implementation.

6.4.1 Building Resilience in Soil

Soil health and fertility management are two important aspects for increasing resilience. Building soil carbon, control of soil erosion and degradation, increasing water holding capacity of soil, balanced fertilisation as per requirement of crop are important interventions for soil health management. Addition of organic manure, crop rotation with legume crops, green manuring and application of *biochar* can build soil carbon. Contour cultivation, land configuration, cover cropping, mulching, strip cropping are some of the practices that could check removal of soil particles from the crop field.

Site-specific nutrient management (SSNM) is a set of location-specific scientific principles for optimally supplying rice with essential nutrients based on specific conditions. The application of SSNM benefits the farmers economically because of lower input costs and possibly higher yields. It also reduces greenhouse gas emissions as more efficient use of nitrogen reduces nitrous oxide emission from the field and indirect emissions during the production of N-fertilizer. Specific nutrient management by using leaf colour chart (LCC) and greenSeeker, use of neem coated urea fertiligation and integrated nutrient management can meet the nutrient requirement of crop and reduce emission of GHG.

Integrated nutrient management deals with the application of organic and inorganic fertilizers, in addition to farmyard manure, vermi-compost, legumes in rotation, and crop residue for sustaining soil health for the long term. Feeding the soil instead of adding fertilizers to the crop without organic inputs is the key point for the long-term sustainability of Indian agriculture

Conservation agriculture is useful to enhance resource use efficiency, provide economic benefits and adaptation to climatic stresses. It improves soil structure and protects the soil against soil erosion and degradation losses by maintaining a permanent soil cover with minimum soil disturbances. It also enhances the fertility status of soil. Crop residues incorporated into soil enhances soil organic matter. Mulches protect the soil surface from extreme temperatures and greatly reduce surface evaporation.

Zero-tillage or no-tillage is designed to generate higher yields while minimizing erosion and reducing costs of inputs and production. Adopters of zero-tillage practices in South Asia, specifically for the rice-wheat system, have reported yields that are greater than or equal to yields obtained through conventional practices. The zero-tillage practice reduces carbon dioxide (CO_2) emissions and saves water use by 25% compared with the conventional methods. The largest CO_2 abatement is achieved through reduced diesel fuel consumption as ploughing or harrowing before planting is not needed. Air pollution is also lessened because the no-till method discourages the burning of crop residue on the field. Zero-tillage can allow farmers to sow wheat sooner after rice harvest, so that the crop escapes the terminal heat stress. Bed planting with residues saves water, enhances farmers' income and also provides resistance to lodging of crops due to unseasonal rains and hailstorm.

Laser land levelling (LLL) is a laser-guided technology used to level fields by removing soil from high points of the field and depositing it in low points of the field. LLL improves crop establishment and enables crops to mature uniformly. In the Indo-Gangetic Plains of India, yields were found to have improved by 8% for rice and wheat; in Cambodia, rice yields improved by 24%. LLL reduces greenhouse gas emissions by saving on energy, reducing cultivation time, and improving input-use efficiency. In a level field, water is distributed evenly, thus, reducing the amount of time and volume of water needed for irrigation. Fertilizer use is more efficient as nutrient runoff from high points to low points in the field is less. Prior to using alternate wetting and drying, LLL avoids too much drying of high points in the field, resulting in a yield penalty during the AWD process.

Fig. 6.6: Laser land levelling

6.4.2 Adapted Cultivars and Cropping Systems

Growing of adapted varieties or breeds tolerant to weather variability and adoption of diversified cropping system can reduce the risks of crop failure due to climate change. Crop management practices like altering planting dates, choice of short duration varieties with tolerance to stressed situations can increase crop yields under adverse situations also. Müller and Elliott (2015) found that adaptive changes in crop management have the potential to increase yields by about 7–15 per cent on average.

Improved and early duration crop varieties adapted to stress, heat, drought, flood and other natural calamities can be promoted and seed production of such varieties can be encouraged. Patterns of drought may need various sets of adaptive forms. To reach deficient downpour conditions, early maturing and drought-tolerant cultivars of green gram (BM 2002-1), chickpea and pigeon pea (BDN-708) were brought on selected farmer's fields in Aurangabad district of Maharashtra (rainfall of 645 millimetres).This provided 20-25 per cent higher yield than the indigenous cultivars. In the same way, drought-tolerant, early maturing cultivars of pigeon pea (AKT-8811) and sorghum (CSH-14) were introduced in the villages of Amravati district, Maharashtra (rainfall of 877 mm).

Local varieties resistant to adverse weather conditions can also be identified and used as climate resistant cultivars along with promotion of seed banks, seed villages and participatory variety selection. A sequential cropping system involving both shallow and deep rooted crops with soil enriching capacity should be preferred. Crop diversification instead of monoculture accommodating a variety of crops can lessen the risk of climate change and weather variability.

6.4.3 Rainwater Harvesting and Recycling

Many rainwater harvesting and conservation technologies like construction of farm ponds, storage structures, restoration of old water harvesting structures, percolation ponds for recharging ground water etc. can be adopted in dryland areas. Inter-row harvesting of rain water in narrow strip cropping with wider intervals, organic mulching, suitable lining in farm ponds, use of anti-evaporants are advisable for restricting surface runoff. Both *in-situ* and *ex-situ* water conservation and recycling of rain water are required for resilient agriculture.

Fig. 6.7: In-situ moisture conservation by BBF System

6.4.4. Water Management

Climate change alters rainfall and water availability patterns and causes water scarcity. Adaptation measures such as water harvesting and storage, access to irrigation, improved irrigation technologies, as well as agronomic practices can enhance water retention capacity of soil. Adaptation to increasing drought conditions and water scarcity can be enabled by enhanced water management in agriculture (HLPE 2015) with water storage and improved access to irrigation water, improved irrigation technologies and techniques such as water harvesting.

Water-smart technologies like a furrow-irrigated raised bed, micro-irrigation, rainwater harvesting structure, cover-crop method, greenhouse cultivation, laser land levelling, reuse of waste water, deficit irrigation and drainage management can support farmers to decrease the effect of variations of climate.

Various technologies based on a precision estimation of crop water needs; groundwater recharge techniques; adoption of scientific water conservation methods; altering the fertilizer and irrigation schedules; cultivating less water requiring varieties; adjusting the planting dates; irrigation scheduling; and adopting zero-tillage may help farmers to get satisfactory crop yields, even in deficit rainfall and warmer years. There are proven technologies like zero tillage, direct seeding of rice, micro irrigation, adoption of improved methods of planting such as broad bed and furrow (BBF) method, land configuration techniques like tied-ridges and tied-furrows are found effective in saving water and increasing crop yields.

Alternate wetting and drying (AWD) is a simple and inexpensive way of reducing water consumption in rice production by 30%, thus, enabling farmers to cut down on production costs without yield penalty. AWD entails periodic draining of the field to a certain threshold, usually 15 cm below the soil surface, and re-flooding. A perforated tube placed in the soil enables the farmer to monitor the water level below the soil surface to determine when to irrigate. The AWD technology has also been proved to effectively mitigate greenhouse gas (GHG) emissions, specifically methane (CH_4), from rice production by 30-70%, without causing a yield reduction. During the dry phases, the methane-producing bacteria are inhibited, thus, setting a condition to reduce GHG emission.

Fig. 6.8: AWD in Rice field

6.4.5 Farm Mechanisation

Farm mechanisation is very critical for timely operations and planting, besides reducing drudgery. Community managed custom hiring centres having sowing/ planting machines, chisel plough, bund former, laser leveller, drum seeder, machine transplanter etc. are required to deal with variable climates.

Dry-seeded rice (DSR) is becoming an attractive option for farmers as it has a much lower labour requirement than manually transplanted rice. In this technology, rice is directly seeded in the main plot with 2- or 4-wheel tractor/ power tiller-drawn seed drills. DSR can be readily adopted by small farmers as well as large farmers, provided that the required machinery is locally available (*e.g.*, through custom hire).

6.4.6 Landscape Management

Many ecosystem services relevant for agricultural production, such as pollination, pest and disease resistance, watershed protection and erosion control, occur over landscape scales. These services are directly related to resilience of agricultural livelihoods through their impacts on reducing environmental risks and improving coping capacities (FAO, 2007; McCarthy et al., 2010). Watershed protection and management, fire management, erosion control, coastal zone management, and pest and disease control are required for landscape management. Landscape-level adaptation will require appropriate institutions and policies to improve coping capacities of communities.

6.4.7 Livestock and Fisheries Interventions

A range of adaptation options like feeding/housing system, production system and breeding can lead to climate resilience. Livestock's adaptive capacity depends on the production system, including choice of species and breeds, the availability/adaptability of alternative feed resources, the accessibility of animals (health/extension services), the type/efficiency of response to outbreaks (surveillance, compensation schemes, etc.) and the household wealth status (ICEM, 2013).

Tolerant breeds in livestock and poultry: Indigenous breeds have unique characters that are adapted to very specific eco-systems across the world. These unique characters are resistant to droughts, thermoregulation, ability to walk long distances, fertility and mothering instincts, ability to ingest and digest low-quality feed, and resistance to diseases. These livestock breeds may not be highly productive in terms of meat or milk production, but are highly adaptive to the unpredictable nature and have low resource footprints.

Feed management: Improvement in feeding systems as an adaptation measure can indirectly improve the efficiency of livestock production. Some feeding methods include altering feeding time or frequency and modification of diet composition, including agroforestry species in the animal diet and training producers in production and conservation of feed for various agro-ecological zones. These measures can decrease the risk from variations of climate by encouraging higher intake or compensating low-feed consumption, decreasing excessive heat load, reducing animal malnutrition and mortality and reducing the feed insecurity during dry seasons respectively.

Adaptation options to declining or variable yields in terms of fisheries technologies and management should be assessed to avoid overexploitation of fisheries or impacting habitats. Practices like diversified and integrated aquaculture systems, water quality monitoring, species selection, selective breeding, genetic improvement, site selection, and improved cage and pond management are required in aquaculture. Management of fish tanks/pond, feed management and water management along with rearing local species are some of the technologies for improvement of fishery in situations of weather variability.

6.4.8 Diversified Forest Ecosystem

Diversified forest ecosystems are more resilient as they are able to cope with stress and recover from damage due to climate change. Healthy ecosystems are more resilient to negative biotic and abiotic influences. Practices like integrated pest management, disease control, forest fire management, employment of reduced impact logging in production forests, limitation of gathering of non-wood forest products or livestock grazing in forests at sustainable levels, and forest law enforcement are found effective in developing a healthy forest ecosystem. Restoration of degraded forest systems is an important strategy for increasing resilience. Thompson et al. (2009) highlight some actions that may be taken to maintain or increase resilience in forests through management and use of biodiversity, at all scales.

6.4.9 Integrated Farming

Integrating different types of production systems (crop, forest, fish and livestock) can reduce the risk of production loss under weather variability. In some regions, switching from crop to mixed crop–livestock or to livestock systems will be a key adaptation strategy (Jones and Thornton, 2009).

6.4.10 Managing Genetic Resources

Crops, livestock, forest trees and aquatic organisms that can survive and produce in future climates will be essential in future production systems (FAO, 2015a, Galluzzi et al., 2011). Genetic resources for food and agriculture are very important for ensuring efficiency, adaptability and resilience in production systems. They underpin the efforts of local communities and researchers to improve the quality and output of food production (FAO, 2015a). *In-situ* and *ex-situ* conservation of domesticated species, their wild relatives and other wild genetic resources need to be promoted for sustainable use in food production system.

6.4.11 Agro-advisory

The development of early warning systems is essential to strengthen proactive decisionmaking at all levels. Analysis of meteorological data, together with information on phenology, soil and agricultural statistics, also allows the provision of near real-time information about the crop forecasting so that timely interventions can be planned and undertaken.

Response farming is an integrative approach; it could be called farming with advisories taken from the technocrats depending on local weather information. The success of response farming, viz., decreased danger and enhanced productivity has already been taken in Tamil Nadu and many other states. Response farming can be a viable choice for climate change adoption strategies, for the variations of climate is not a sudden one. The main cause for the success of response farming is because of both location and time-specific technologies. It is time to take forward the success of response farming to the entire farming community.

6.4.12 Diversification in on-farm and off-farm

Non-agriculture-based livelihoods can build resilience among agricultural populations due to projected population growth patterns as well as potential climate change impacts. Diversification, Both on-farm with increased number of varieties, species and breeds, including through mixed systems such as crop/livestock, crop/fish or processing products, and off-farm, by getting a non-agricultural job, is an important element of climate change adaptation (Thornton and Herrero, 2014).

6.5 National Programmes for Climate Change Adaptation in India

The convergence of various policy programmes and sectoral plans has been undertaken by the Government of India to ensure synergy and effective utilisation of existing resources. The National Mission of Sustainable Agriculture was implemented in 2010 under the National Action Plan on Climate Change (NAPCC) to promote the judicious management of available resources and this was one of the eight missions under NAPCC.

The Pradhan Mantri Krishi Sinchayee Yojana (PMKSY) was launched in 2015 to address the issues of water resources and provide a permanent solution that envisages Per Drop More Crop, by promoting micro / drip irrigation for the conservation of maximum water.

The Paramparagat Krishi Vikas Yojana mission was executed to extensively leverage adaptation of climate-smart practices and technologies in conjunction with the Indian Council of Agricultural Research and state governments of India.

To mitigate climate extreme actions, Green India Mission was launched by the GOI in 2014 under the umbrella of NAPCC with the primary objective of protecting, restoring and enhancing India's diminishing forest covers, thereby reducing the deleterious effects of climate change.

To protect the soil health, GOI has launched the Soil Health Card scheme with the main objective of analysing cluster soil samples and advocating farmers regarding their land fertility status. Additionally, Neem-coated Urea was also introduced to minimise the excess addition of urea fertilizers, thereby protecting soil health and supplying plant nitrogen.

To encourage farmers with more income benefit and ecosystem protection, programmes such as the National Project on Organic Farming and National Agroforestry Policy was introduced in 2004 and 2014 respectively. These policies are aimed at supplying plant nutrients in the form of organic amendments, soil carbon stock improvement, and soil protection from erosion loss.

Andhra Pradesh, Himachal Pradesh, Sikkim, etc., have already initiated several programmes to adopt and promote organic farming practices on a wider scale. Recently, GOI announced Sikkim as organic state.

The ICAR, through its network research Institutes, state agriculture universities and all line departments is implementing agriculture contingency plans in about 650 districts of India towards climate change preparedness for the last seven years.

These models are taken forward to SAARC countries towards adaptation to climate change impacts like floods, cyclones, droughts, and heat waves and seawater intrusion. ICAR has established climate-resilient villages across India in 151 districts, which are replicated by the state governments towards the overall objective of building carbon positive villages.

The Mahatma Gandhi National Rural Employment Guarantee Act (MGNREGA) was initiated in 2005 with the key objective of "enhancing employment opportunities, additionally, providing economic security and protecting the environment".

References

FAO (2012): What is Conservation Agriculture? Available at: http://www.fao.org/ag/ca/1a.html (accessed 21, December 2017).

FAO, 2021. Climate Resilient Practices: Typology and guiding material for risk screening

FAO. 2007. The State of Food and Agriculture 2007. Paying farmers for environmental services. Rome

FAO. 2012a. Building resilience for adaptation to climate change in the fisheries and aquaculture sector [online]. [Cited 23 February 2021]. http://www.fao.org/3/i3084e/i3084e08. pdf

FAO. 2013b. Climate-Smart Agriculture Sourcebook. Rome, FAO. 558 pp. (also available at: http://www.fao.org/3/a-i3325e.pdf)

FAO. 2015a. Coping with climate change – the roles of genetic resources for food and agriculture. Rome (available at http://www.fao.org/3/a-i3866e.pdf)

Galluzzi, G., Duijvendijk, C. van., Collette, L., Azzu, N. & Hodgkin, T., eds. 2011. Biodiversity for food and agriculture: contributing to food security and sustainability in a changing world. Outcomes of an Expert Workshop held by FAO and the Platform on Agro-biodiversity Research from 14-16 April 2010, Rome, Italy. Rome, FAO/PAR (available at http://www.fao.org/3/a-i1980e.pdf).

HLPE. 2015. Water for food security and nutrition. A report by the High Level Panel of Experts on Food Security and Nutrition of the Committee on World Food Security, Rome.

ICEM (International Centre for Environmental Management). 2013. USAID Mekong ARCC climate change impact and adaptation on livestock. Prepared for the United States Agency for International Development by ICEM

Jones, P.G. & Thornton, P.K. 2009. Croppers to livestock keepers: livelihood transitions to 2050 in Africa due to climate change. Environmental Science & Policy, 12(4): 427–437.

McCarthy, M., Best, M. & Betts, R.2010. Climate change in cities due to global warming and urban effects. Geophysical Research Letters, 37(9)

Müller, C. & Elliott, J. 2015. The Global Gridded Crop Model inter comparison: approaches, insights and caveats for modelling climate change impacts on agriculture at the global scale. In A. Elbehri, ed. Climate change and food systems: global assessments and implications for food security and trade. Rome, FAO.

NICRA & CRIDA: Climate Resilient Agriculture through adaptation and mitigation strategies

Prabhakar, M.2022: National Innovations in Climate Resilient Agriculture (NICRA); A Flagship Project of ICAR Addressing Climate Change in Agriculture: In Strategies for Climate Resilient Farming (CRIDA & MANAGE, Hyderabad)

Thompson, I., Mackey, B., NcNulty, S. & Mosseler, A. 2009. Forest resilience, biodiversity, and climate change. A synthesis of the biodiversity /resilience/stability relationship in forest ecosystems. Technical Series No. 43. Montreal, Canada, Secretariat of the Convention on Biological Diversity

Thornton, P.K. & Herrero, M. 2014. Climate change adaptation in mixed crop–livestock systems in developing countries. Global Food Security, 3(2): 99–107

Tripathy, R & Bisen, J,2019: Climate Resilient Agriculture Technologies for Future (Training Manual): ICAR-NRRI, Cuttack

7

Basics of Climate Smart Agriculture

7.1 Introduction on Climate Smart Agriculture (CSA)

Extreme weather conditions like rainstorms, high wind pressures, high temperatures, flood and drought situations have profound influence on agriculture. Climate and agriculture are strongly interrelated universal processes and thus variations in climate influence agricultural activities (IPCC, 2007). The effects of climate change are seen in form of global warming, unseasonal and excess rainfall, and frequent natural calamities which affects agricultural production. At the same time emission of GHG to the atmosphere is a major source of climate change. The impact of climate change on crop yield, dairy and other livestock production including fishery will be felt in reduced production and consumption. Climate change will effect in additional increase in prices for main agricultural food crops such as rice, wheat, corn and soybeans (IFPRI, 2009). In order to face these challenges, Climate Smart Agriculture (CSA) has been suggested by FAO for National Food Security and Development Goals (Fig.7.1).

Climate-smart agriculture is an innovative approach that can make the agriculture sectors more productive and sustainable with better ability to contributing towards climate change adaptation and mitigation. FAO defined climate-smart agriculture (CSA) as the three dimensions of sustainable development (economic, social and environmental) by jointly addressing food security and climate challenges. Agriculture that sustainably increases productivity and resilience (to climate change), reduces/removes greenhouse gases (mitigation), and enhances achievement of national food security and development goals (FAO, 2011).

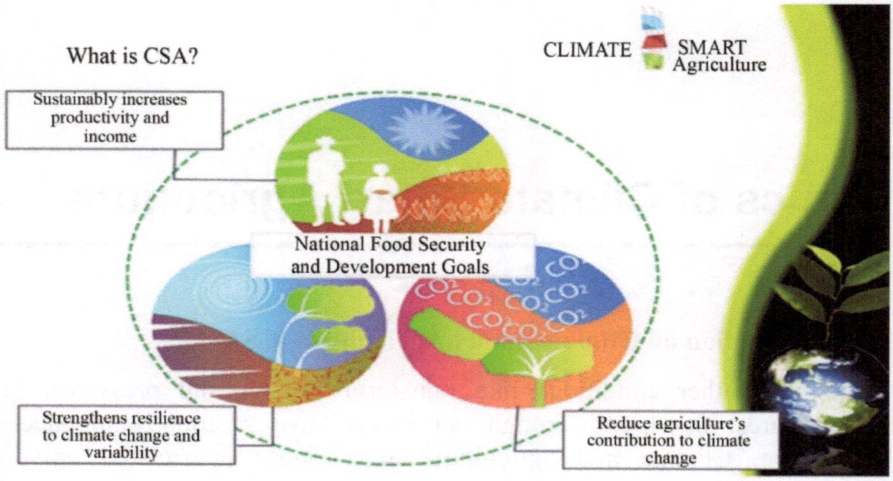

Adapted from FAO

Fig. 7.1: National Food Security and Development Goals

The World Bank Group defines climate –smart agriculture (CSA) as an integrated approach to managing cropland, livestock, forests and fisheries that address the interlinked challenges of food security and climate change. Thus climate smart agriculture is regarded as a new approach to support farmers to achieve sustainable agricultural production and land management with ability of the sector to adapt and mitigate the adverse effects of climate change. CSA is a technique to improve the scientific rule and investment setting to attain sustainable agricultural progress to ensure the food availability under climate change (FAO, 2012). It can ensure food availability and conservation of natural resources. It relies on adaptation process which involves management of risks generated by climate change, including variability. CSA is based on three pillars, such as (i) increase in raising income, productivity and food security, (ii) adaptation and resilience to climate change, and (iii) reducing GHG emissions, where possible (Fig.7.2). With the integrated policies for adaptation, mitigation and ensuring food security the outputs would be increased productivity, increased net return, improved input use efficiency, reduction in emissions of GHG, increased resilience to climate change and increased gender and social inclusions (Fig 7.3).

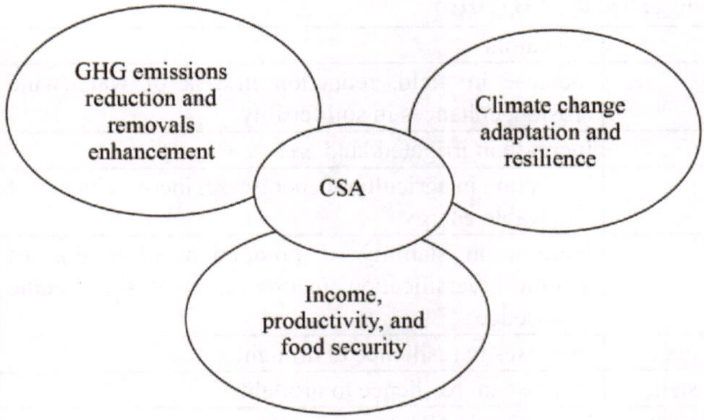

Fig. 7.2: Three pillars of CSA

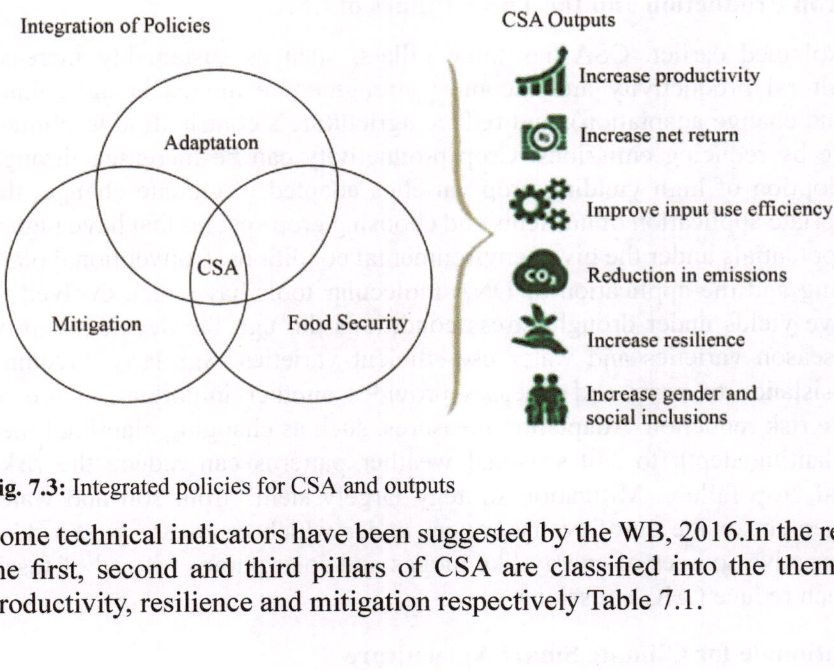

Fig. 7.3: Integrated policies for CSA and outputs

Some technical indicators have been suggested by the WB, 2016.In the report, the first, second and third pillars of CSA are classified into the themes of productivity, resilience and mitigation respectively Table 7.1.

Table 7.1: Indicators of CSA suggested by WB (2016)

Theme/pillar	Sub-theme	Indicators
Productivity	Crop system	Increase in yield, reduction in risk of water/wind erosion, enhances in soil fertility
	Water use	Increase in irrigated land area
	Energy	Reduction in agricultural energy use, increase in use of renewable energy
Resilience	Robustness	Increase in stability of production, promotion of income diversification, incorporation of site-specific knowledge
	Cropping system	Increases in resilience to drought
	Livestock system	Increase in resilience to drought
Mitigation	Emission intensity	Reduction in GHG emissions
	Sequesters carbon	Increase in carbon sequestration

7.2 Crop Production and the Three Pillars of CSA

As explained earlier, CSA has three pillars, such as sustainably increase agricultural productivity and incomes; strengthen resilience in agriculture (climate change adaptation); and reduce agriculture's contribution to climate change by reducing emissions. Crop productivity can be increased through the adoption of high yielding crop varieties adapted to climate change, the appropriate application of nutrients and choosing crop species that have higher yield potentials under the given environmental conditions. Conventional plant breeding and the application of DNA molecular tools have been evolved to improve yields under drought stress conditions through the development of short-season varieties and water use-efficient varieties. Similarly, breeding for resistance to pests and diseases provides another important source of climate risk reduction. Adaptation measures, such as changing planting times and planting depth to suit seasonal weather patterns can reduce the risks of total crop failure. Mitigation strategy largely stems from soil and water management or the agroforestry system under which crops are grown. Use of alternative sources of energy like biogas and solar pumps instead of fossil fuels can reduce GHG emissions.

7.3 Rationale for Climate Smart Agriculture

The Food and Agriculture Organisation (FAO) estimates that feeding the world population of nearly 9.8 billion by 2050 will require a 60 per cent increase in total agricultural production. Climate change is already negatively impacting agricultural production globally and locally as agriculture currently generates 19–29% of total GHG emissions. Unpredictable weather patterns, shorter growing seasons, droughts, extreme temperatures, and increased exposure to pests and crop diseases pose serious problems to smallholder farmers around the world. Shrinking land and other resources for agriculture and their unsustainable management practices have made the agriculture unsustainable. To meet this challenge, the agriculture sectors must improve their sustainability and adapt to the impacts of climate change to ensure food security for all. Climate Smart Agriculture would be a major instrument to achieve this. It seeks to reduce trade-offs and promote synergies to make crop and livestock systems, forestry, and fisheries and aquaculture more productive and more sustainable by reducing the impact of climate change.

7.4 Objectives of CSA

There are three broad objectives of CSA.

(i) **Sustainably maintain and increase productivity**: In order to ensure food and nutrition security our diets need to be balanced with carbohydrate, protein, fat, minerals, vitamins and micronutrients. Our future demand for food will include animal products such as milk, egg and meat/fish. Therefore there is a need for sustainable production of diversified food.

(ii) **Adaptation and resilience**: Global warming, rising sea levels, extreme weather situations due to climate change have affected agricultural production in most of the countries. In order to ensure resilience of agriculture systems, it is inevitable that the practices are adapted to climate change resilient.

(iii) **Mitigation:** Rising levels of GHG causing climate change have to be reduced to the acceptable level. Currently agriculture (and related sectors) contributes about a quarter of human-induced GHG emissions, so there is a lot of potential to reduce these emissions, and to mitigate other detrimental effects of agriculture on the environment. Reducing GHG by some mitigating measure is desirable to reduce the impact of climate change. Another important aspect of mitigation is the uptake of carbon in plants and soils, which can help to reduce the concentration of CO_2 in our atmosphere.

The three broad objectives of CSA are displayed in Fig.7.4.

Fig. 7.4: Broad objectives of CSA (FAO, 2017)

7.5 Approaches to CSA

Climate-smart agriculture is an approach that helps guide actions needed to transform and reorient agricultural systems to effectively support development and ensure food security in a changing climate (FAO). CSA combines various sustainable methods to tackle the specific climate risks of a farming community. It is an integrated approach to managing different agricultural sectors to address interlinked challenges of food security and climate change.

The first step is to assess the particular climate risks by using variety of tools to assess the nature climate risk and vulnerability of a landscape, taking the local ecosystems and the specific crop into account. Finding the right combination to manage a specific climate challenges and to build resilience to future impacts, makes climate-smart agriculture "smart." New climate risks require changes in agricultural technologies and approaches to improve the food insecurity and poverty. Addressing the global challenges of climate change requires generation, adaptation and use of new knowledge. This encompasses the need to: link more effectively and responsively to domestic and international markets (food, feed, fibres, etc. and/or carbon); reduce the vulnerability and enhance the voice of the rural poor; promote environmental conservation; build linkages between farmers and other agencies; and institutional and organisational development to support the bargaining position of farmers. Climate smart agriculture requires investments on (i) managing climate risks, (ii) understanding and planning for adaptive measures, and (iii) exploring options for reducing emissions of greenhouse gases, where possible.

7.6 Elements of CSA

Different elements which can be integrated in climate-smart agricultural approaches include: (i) management of farms, crops, livestock, aquaculture and capture fisheries to manage resources better, produce more with less while increasing resilience, (ii) ecosystem and landscape management to conserve ecosystem services that are key to increase at the same time resource efficiency and resilience, (iii) enabling the farmers to manage the risks and impacts of climate change and undertake mitigation actions, and (iv) changes in the wider food system including demand-side measures and value chain interventions that enhance the benefits of climate smart agriculture.

(i) **Management of farms, crops, livestock, aquaculture and capture fisheries**: Options for management of crops include switching varieties or species, changing cropping calendars, and nutrient management such as micro-dosing, mulching or organic fertilizers application etc. Options for livestock management include improving the quality of pastures and feed, grazing management, changing herd management, and specific responses to heat stress. In fisheries sector, introduction of local species, pond management, water and feed management are possible options for climate resilience. Overall farm-management options include diversification of production, integrated crop-livestock systems, agroforestry, restoring organic soils, limiting soil erosion, increasing energy efficiency, use of biomass fuels, integrated pest management, and enhancing management of water resources and irrigation.

(ii) **Landscape or ecosystem management:** CSA also encourages looking at agricultural systems in the context of larger landscapes and ecosystems, so as to better understand the inter-linkages between agricultural production and ecosystem services within and external to agro-ecosystems. Water resource management and land-use change in food security, adaptation and mitigation across landscapes are important elements.

(iii) **Services for farmers and land managers**: In order to increasing adaptive capacity of farmers, required services include climate information services, such as seasonal forecasts or early-warning systems, advisory services that link climate information to agricultural decisions, and financial services such as credit and insurance. Social protection as well as new index-based weather insurance products can increase the ability of smallholders to invest in agriculture despite increasing climate variability.

Fig. 7.5: Use of leaf colour chart to judge the nitrogen content required for crops

(iv) Changes in the wider food system: Agricultural production not the only focus on adaptation and mitigation actions that support food security and livelihoods. Across the value chain, innovations in harvesting, storage, transport, primary and secondary processing, retail and consumer activities are essential elements of the enabling and incentivising environment needed for climate smart agriculture.

7.7 Strategies for CSA

The impacts of climate change can be reduced by adopting some strategies for land use management, improved water management, resource conservation, crop diversification, use to tolerant/resistant varieties to adverse weather, better weather forecasting, and improved pest management etc.

7.7.1 Efficient resource management

Resource management is very critical in CSA as there is substantial food loss throughout the food chain. Therefore energy use efficiency has to be increased by reducing food loss as about 38% of annual energy is utilised in the food chain.

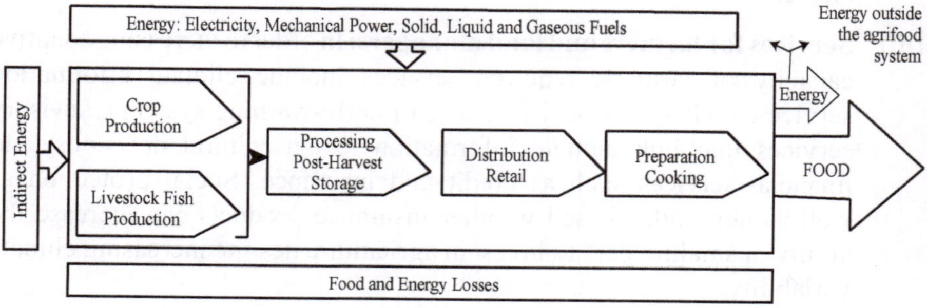

Fig. 7.5: Food and energy losses (FAO)

7.7.2 Integrated Renewable Energy Technologies for Farming Systems

In order to create a stable change to energy smart and proficient food systems, many technologies are required to be adopted. Such technologies may include wind mills, solar panels, biogas production, ethanol extraction as a bio-fuel, bio-energy operated water pumps, renewable energy-powered vehicles, solar-wing electricity production etc.

7.7.3 Availability of Technical Knowledge of Farmers

Conventional environmental knowledge of people improved and developed in the test of time could give different ideas and feasible options for adaptation to climate change. Indigenous technologies with the farmers may be validated and used in climate smart agriculture.

7.7.4 Resource conserving technologies

The resource conserving technologies (RCTs) are required to reduce production cost and saving resources. The RCTs such as zero tillage systems (ZT) or no-tillage/ minimum tillage is a cultivation system in which seeds are directly sown in uncultivated soil. The seeds are sown in untilled soil by creating thin channels that only have adequate depth and width to attain suitable seed coverage. Remaining soil is left as if no tillage is done with it (Derpsch et al., 2010).Minimum disturbance of soil encourages retention of more organic matter in the shape of stored carbon for which it does not contribute much to global warming. Therefore no-till has much share to greenhouse gas alleviation by improved carbon sequestration in the soil (Zingore, et al. 2010).

7.7.5 Crops Genetic Modification

Availability of new crop varieties that are tolerant to heat, drought and salinity can reduce the risks of adverse climate situation. Genetic diversity of the seed structure and seed composition has been recognized as a very effective defence against plant disease and pest attack and risks of climate. Genetic modification has doubled the vegetable production in the last 25 years and due to this, worldwide trade in vegetables is much higher in volume than that of cereals.

7.7.6 Land use Management

Land use management practices like shifting production between livestock and crops, reducing use of chemical fertilisers and pesticides, adjusting time of sowing and harvesting of crops for altering length of seasons of growth and levels of changing heat and humidity periods are some of the options to minimise the risk of climate change. Adjusting the time of sowing/planting can regulate the length of the growing season for the better adaptation to the altering environment.

7.7.7 Cropping Season Variation

Changing the sowing and planting dates can coincide the crop growth period with the wet period which can avoid intense weather events in the crop growing season. Changed cultivation systems include improving the better cultivars and enhancing the intensity of farming in various crops. Farmers will have to manage the changes in different hydrological regimes by adopting changed crop rotations (Pathak et al., 2012).

7.7.8 Crops Relocation

Temperature rise may affect the quality of crops such as aromatic rice and tea. Other areas with cooler climate might be suitable for these crops for assured quality. Such areas must be identified to accommodate such crops, which are generally prone to climate change / variability.

7.7.9 Efficient Pest Management

Crop pests and diseases are favoured by some specific temperature and rainfall pattern. Hence variations of temperature and rainfall unpredictably influence pest and disease incidence due to the effect of climate change that will inherently affect the relationship of pest / weed and natural enemies. Integrated Pest Management (IPM) tactics involving use of resistant/tolerant crop varieties, cultural control, biological control, mechanical control and adoption of appropriate crop production techniques need to be adopted.

7.7.10. Carbon Sequestration

Researchers have advised farmers to promote carbon sequestration, i.e., the natural and deliberate processes by which CO_2 is removed from the atmosphere or diverted from emission sources and stored in the ocean, terrestrial environments (vegetation, soils and sediments) and geological formations.

7.8 Methods of CSA

Agro-ecological principles are required to be followed in deciding the methods of climate smart agriculture. These principles are:

- Increasing the biomass in soil to optimizing organic matter decomposition and nutrient cycling;
- Strengthening the resilience of agricultural systems through enhancement of functional biodiversity, by creating habitats for natural enemies of pests;
- Managing soil organic matter to increase biological activity and providing a favourable condition for plant growth;

- Minimizing losses of energy, water, nutrients and genetic resources by enhancing conservation and regeneration of soil and water resources and agro-biodiversity.
- Diversifying species and genetic resources in agro-ecosystem; and
- Enhancing biological interactions and synergies among the components of agro-biodiversity, thereby promoting key ecological processes and services.

Various methods suggest enhancing the ability of the farmers to maximise the agricultural production even under adverse climatic situations. The potential CSA methods are outlined below.

7.8.1 Use of Quality Seeds and Planting Materials of Crops/Varieties of Well-adapted Crops and Varieties to Weather Variability

Quality seeds and planting materials of well-adapted varieties are very crucial for CSA. Plant breeding efforts involve multi-locational trials to develop crop varieties that are resistant to climate-related adverse situations. Resistance to climate risks like drought, salinity, stress, high temperature and flooding are the most common climate-related traits for which crop varieties are evolved through breeding. Well-adapted crop varieties to such climate risks are released for ultimate goal of ensuring farmers have access to quality seeds and planting materials. The farmers also obtain seeds from multiple unregulated sources which include saving seeds from their own harvests, purchasing seeds from local markets and exchanging seeds with family members and neighbours. Such informal systems predominate in developing countries where crop production systems are the most vulnerable to extreme weather events. Such informal seed supply system has also to be by community-based seed production and distribution channels. But quality assurance is very important in both formal and informal seed distribution system. There is also a need for quality assurance in informal seed distribution system by encouraging seed banks, seed village programme etc.

7.8.2 Biodiversity Management

Crop diversity is especially important in a climate-smart approach because it contributes to pest and disease management and crop yield. Enhancing on-farm biodiversity also provides other environmental services, including pollination, that are essential to farmers and society as a whole. In a cropping system, greater diversity of crops and other living organisms is an important criterion for ensuring farm resilience, economic stability and profitability. Diversification is especially important in a climate smart approach as it contributes to ecological pest management and crop yield. Growing multipurpose crop varieties can provide food, bio-fuel, feed, fibre to serve the multiple needs and provide climate resilience. This is needed to ensure the crops selected are appropriate for the specific conditions of each farm system which do not compete for nutrients, water and light. These options include mixed cropping, relay cropping, inter-cropping, crop rotation, cover crops etc.

Fig. 7.5: Mixed cropping Fig.7.6 Inter-cropping

7.8.3 Integrated Pest Management (IPM)

Crops are usually affected by various insects, disease causing pathogens, weeds and other harmful organisms in the crop ecosystem. At the same time there are many beneficial organisms which provide a natural check on the harmful organisms. Indiscriminate use of synthetic chemicals have upset the natural ecosystem and invited more pest attack. Sole reliance on toxic chemicals for pest control may upset the pest-defender interaction and result pest resurgence and endemics. In order to avoid such environmental problems, integrated pest management is recommended. The integrated pest management (IPM) is a broad ecological approach which utilises all available tools of pest management judiciously to manage the pest population within the economic injury level. The host-plant resistance by using resistant/tolerant crop varieties, cultural control, physical& mechanical control, biological control and need based chemical control at a stage when the pest population exceeds the economic threshold level (ETL) are usual methods of IPM (Fig.7.7). Farmer Field Schools, which facilitate learning by doing and experimentation of different management options by farmers, educate and train the farmers to adopt IPM technologies. As the climate changes, national regulatory, policy and institutional frameworks must be strengthened to enable the adoption of integrated pest management practices on farms and in rural communities.

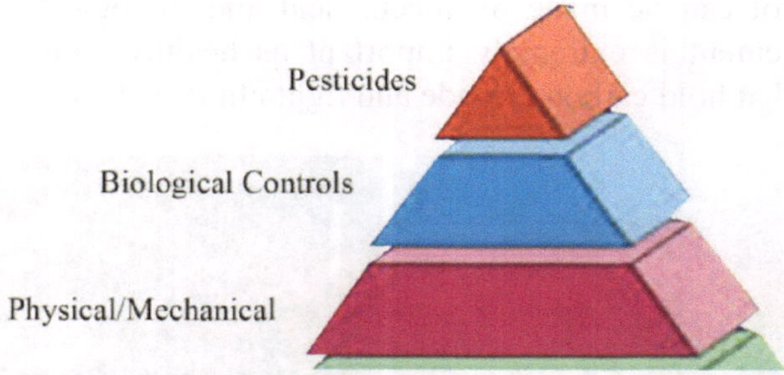

Fig. 7.7: Integrated pest management tactics

7.8.4 Improved Water Use and Management

Considering the future water scarcity for agriculture, increasing water use efficiency and rainwater management have become very crucial. Harvesting rainwater can be done by digging farm ponds, constructing water harvesting structures and check dams for conservation of rain water. Water management (rain water or irrigation water) is very important in areas of climate risk. Where water is a limiting factor, improving water management can be achieved through measures that conserve soil and water; and/or with deficit irrigation that can maximize crop yields per volume of water applied; and/or more efficient irrigation technologies that can reduce unproductive evaporation losses. In areas where water is limiting and energy cost is very high, the best option would be to use solar pumps with micro-irrigation system.

7.8.5 Sustainable Soil and Land Management for Increased Crop Productivity

Conservation agriculture is an approach that combines limiting soil disturbance to a minimum, maintaining soil cover and diversifying crop production. Conservation agriculture provides a strategic entry point for climate change adaptation. Keeping the soil covered by cover crops reduces moisture loss, stabilizes soil temperature, reduces erosion by water and wind, restores soil carbon through the decomposition of crop residues, and provides food for beneficial soil organisms. Crop rotation, cover cropping, minimum tillage, organic recycling in soil by incorporating crop residues, mulching, addition of organic matters etc. are very helpful for soil and land management. The organic matter incorporated by soil micro-fauna into the soil improves soil structure and water storage capacity, which in turn helps plants survive longer during periods of drought. Minimum tillage is a long-term management approach for increasing the amount of carbon stored in the soil. When soil slowly rebuilds, it grows and stores more soil organic matter and water, thus improving ecosystem functions and services to make it more climate resilient.

Erosion control can be made by mechanical and biological measures.Soil health management is extremely important as healthy soils are important carbon sinks that hold carbon dioxide and fight climate change.

Fig. 7.8: Contour cultivation on sloppy lands

Fig. 7.9: Mulching in CSA

7.8.6 Sustainable Mechanization

Use of appropriate machinery can increases efficiency in the various production and processing operations and in the production, extraction and transport of agricultural inputs, including coal and oil. Using smaller tractors, making fewer passes across the field and reducing working hours, can reduce carbon dioxide emissions, minimize soil disturbance, and reduce soil erosion and degradation that are common in tillage-based crop systems. Farm equipment required for precision agriculture along with controlled release and deep placement technologies will enable to match the production inputs with plant needs. Agriculture machinery like solar pumps can reduce the use of fossil fuels and emission of GHG. All these measures can reduce the emission of GHG.

Fig. 7.11: Machinery used in CSA

7.8.7 Planting Shade Trees

Planting shade trees can help protect the farm from excessive sun, harsh winds, and strong showers. Trees planted as shelter belts can resist wind erosion Excessive shade, however, can contribute to a more humid microclimate on the farm, and with cocoa, for example, excessive humidity creates more favourable conditions for certain fungal diseases.

7.9 Climate Smart Action Plan

7.9.1 Climate Smart Cropping Systems

There are many options for climate change adaptation and mitigation across different cropping systems depending upon types of stress, the farmers' coping and adaptive mechanisms, and the degree to which each climate factor affects yields. The sustainable intensification of crop production guides decisions on how to overcome inefficiencies that result in yield and productivity gaps and minimize the negative environmental and social impacts of production. It calls for farming practices that uses good quality seeds and planting materials of well-adapted varieties; cultivation of a varied range of crop species and varieties in associations, intercrops or rotations; control of pests through integrated pest management practices ; and the adoption of conservation agriculture and sustainable mechanization to maintain the soil health and improve water use efficiency to achieve the highest possible crop yields within the carrying capacity of the ecosystem.

7.9.2 Climate Smart Forestry

Apart from providing livelihoods for the rural communities by supplying food, fodder, fuel, medicinal herbs and raw materials for industry and construction, forests act as buffers against shocks and provide ecosystem services that influence agricultural production. They protect water and soil resources, assist in soil development and enhance soil fertility, regulate climate, and provide habitats for wild pollinators and predators of agricultural pests. Forests help control climate by removing large amounts of carbon dioxide from the atmosphere and acting as carbon sinks, storing the carbon in the soil and biomass. Air temperature, solar radiation, rainfall, and atmospheric carbon dioxide concentrations are major drivers of forest productivity and forest dynamics. They are also a source of carbon dioxide emission in that they release carbon dioxide into the atmosphere through respiration. Since deforestation and forest degradation account for an estimated 17 per cent of global greenhouse gas emissions, reducing emissions from deforestation and forest degradation including sustainable management of forests and enhancement of forest carbon stocks will play an important role to combat climate change.

7.9.3 Climate Smart Livestock Production

Livestock is an important component of farming system. In mixed systems, livestock consume crop residues and by-products and their manure is used to provide nutrients to crops. The contribution of livestock for rural livelihoods goes far beyond agriculture production and food security; it directly supports social welfare, education and human health. Careful livestock management is required to maximize the range of services they provide and reduce the impacts of climate change. There are many synergistic options for both adaptation and mitigation that can be implemented in the livestock sector, including shifts in species and breeds, improved feed management, sustainable grazing practices, silvo-pastoralism and on and off-farm diversification etc.

7.9.4 Climate Smart Fishery

Inland fishery and captured fishery are affected by climate change.Population growth is increasing the demand for food, and unsustainable fishing practices have demanded more production from marine fisheries. Aquaculture will have to bridge this gap between supply and demand. There are numerous options for supporting such growth in a climate-smart and sustainable manner, including sustainable water management practices; selective breeding and genetic improvements; improved feed management; strengthened emergency procedures and biosecurity measures.

7.9.5 Integrated Production System

Integrated production systems are also called integrated farming system where the by-product of one component is used as input in the system. Examples of integrated production systems include agroforestry and crop-livestock, rice-fish, fish-livestock and food-energy systems. The diversification of resources and incomes associated with integrated production system offers producers a greater number of risk management strategies and options to adapt to the impacts of climate change. Also, the greenhouse gas emission intensities of integrated systems are typically lower than those of specialized systems. The sustainable intensification of production on integrated agricultural systems requires a better understanding of the impacts of climate changes and climate variability on these systems.

7.9.6 Water Management and Climate Change

Climate change affects the water cycle. Agriculture is affected by increased evaporative demand due to temperature rise. Changes in the amount of rainfall and variations in rainfall patterns will affect the two sources of water for irrigation: river flow and groundwater recharge. Agriculture is responsible for 70 per cent of freshwater withdrawal globally for crop production. Rainwater harvesting, use of marginal water and wastewater resources, water use efficiency measures and watershed management procedures are some examples of options identified for effective water management to address the effects of climate change.

7.9.7 Sustainable Soil and Land Management

In cropping, grazing and forest systems, climate change affects soil health and plant growth. Erratic rainfall and more frequent and severe periods of drought will reduce the capacity of soils to supply water and nutrients to plants. Higher evaporation and transpiration rates will lead to increased erosion, reduced groundwater recharge and reduced soil moisture for plant growth, and occurrence of soil salinization. Higher soil temperature would favour the rate of mineralisation of soil organic matter that will reduce the potential of soil for better plant growth. GHG emissions from soil can be reduced by proper land use management, land use change, vegetation cover, soil management and carbon sequestration. Sustainable soil and land management initiatives that build up soil organic matter can provide better soil ecosystem services and climate resilience.

7.9.8 Genetic Resources for Food & Agriculture

Selective breeding and domestication, combined with natural selection, have led to the development of numerous and diverse varieties, breeds, stocks and strains of plant, animals and microorganisms. Such genetic diversity is lost due to agricultural intensification and use of agro-chemicals. As different plants, animals and microorganisms have different capacities to survive or adapt to shocks and changes, safeguarding and sustainably managing the diversity of genetic resources is critical for climate change adaptation and mitigation strategies. Characters like water stress tolerance, drought resistance, or resistance to pest or diseases, may be included in breeding or conservation strategies designed to build resilience to climate change. The indigenous and wild crop varieties should also be preserved as these are better adapted to climate change and variability.

7.9.9 Energy-smart Food

Productions of food and post-harvest activities have been energy-intensive. This has resulted in greenhouse gas emission also. The agri-food sector is estimated to contribute about 30 per cent of global greenhouse gas emissions. A sustainable approach must focus on the water-energy-food nexus, and address trade-offs and capitalize on synergies in the use of these resources. Food losses occur at all stages of the supply chain and global annual food losses amounts to around 38 per cent of the energy consumed by the whole food chain. This could be compensated through the sale of biogas produced on farms to local households, or through the generation of electricity from agricultural residues that can be fed into the national energy grid. To make food systems more environmentally sustainable and climate-smart, it is important to examine the food system's carbon footprint, and where possible, identify more efficient, less carbon-intensive approaches.

References

FAO. 2011. Save and grow: a policymaker's guide to the sustainable intensification of small holder crop production. Rome.

FAO, MICCA Series No. 4. Rome, FAO. NHB .2011. National Horticulture Database, 2011, Ministry of Agriculture, Govt. of India, Gurgaon, pp: 278.

FAO. 2012. The Sahel crisis: executive brief. March. Rome. Fisheries

FAO, 2019: Climate-Smart Agriculture in action: from Concepts to Investments. Dedicated Training for Staff of the Islamic Development Bank

IFPRI. 2009. Climate Change: Impact on Agriculture and Costs of Adaptation. International Food Policy Research Institute, Washington.

Pathak H, Aggarwal PK, Singh SD. (Editors). 2012. Climate Change Impact, Adaptation and Mitigation in Agriculture: Methodology for Assessment and Applications. Indian Agricultural Research Institute, New Delhi. Pp: 302.

Zingore S. 2010. SOC sequestration in farming systems in Africa: Potential, opportunities and chal-lenges, In: 2nd Meeting of the Round Table on Organic Agriculture and Climate Change, RTOACC, 2010.

8

Broad Strategies for Climate Smart Agriculture

8.1 Need for climate-smart agriculture strategy

Climate change has long-lasting and potentially detrimental consequences on the agriculture sector. Changes in weather events cause perturbations across agricultural ecosystems and have impact on farm productivity. Adopting the climate-smart agriculture approach has thus, become inevitable in order to build resilience in agriculture systems and cope up with climate change. With a view to implementing climate-smart agricultural practices to ensure climate resilience, certain broad strategies have to be adopted. The CSA approach is designed to identify and operationalize sustainable agricultural development within the explicit parameters of climate change (Climate Smart Agriculture Sourcebook.FAO.2013). It integrates the three dimensions of sustainable development (economic, social and environmental) by jointly addressing food security and climate challenges.

Various strategies for CSA are general plans of action intended to address the adverse effects of climate change in the long run. While preparing the strategies, the impacts of climate change on agriculture are kept in mind.

8.2 Climate smart strategies

There are three broad strategies for climate smart agriculture. Firstly, a set of strategies are adopted for increasing crop yields under weather variability. Secondly, a set of strategies are adopted to make the agriculture system adapted to climate change. Thirdly, mitigation strategies are adopted for reducing the GHG emissions and managing the risks of climate change. The CSA options integrate innovative and traditional technologies, practices and services that are relevant for particular location and reduce the effect of climate change. Substantial reduction in GHG emission over the years can reduce the occurrence of climatic variability, increase prospects for effective adaptation, reduce the costs and challenges of mitigation and contribute to climate-resilient pathways for sustainable agriculture.

8.2.1 Management Strategies for Increasing Crop Yields under Weather Variability

Agricultural technologies and practices such as minimum tillage, different methods of crop establishment, nutrient and irrigation management and residue management can improve crop yield, nutrient and water use efficiency, and reduce greenhouse gas (GHG) emissions from agricultural activities. The two important farm management strategies that employ climate-smart methods to boost agricultural productivity sustainably are given below.

(i) Sustainable soil management

Soil management is crucial for minimising the adverse impacts of climate change. Preserving and maintaining organic matter in soil can enable soils to be resilient to the extreme weather events. A healthy equilibrium of minerals, organic matter, living organisms, air and water in soil will ensure that essential nutrients are available to the crops as per requirement. Use of soil test based nutrients and critical inputs with user-friendly composting techniques, farm management systems can enable the farming to adapt to the constant climate changes and successfully retain the natural health of soil.

Preventing and mitigating land degradation: Degraded soils are vulnerable due to losses of soil organic matter and soil biodiversity, soil compaction and increased rates of soil erosion and landslides. It is estimated that overall, land use and land use changes account for around 31 per cent of total human-induced GHG emissions (Scherr and Sthapit, 2009b). Land degradation is caused by physical, biological and chemical degradation. Physical degradation is caused by water, wind, surface crusting, hard pan development and waterlogging. Biological degradation results from depletion of soil organic matter and loss of flora and fauna. Chemical degradation is caused by soil acidification, salinity, nutrient depletion, pollution from indiscriminate use of fertilisers and pesticides including industrial wastes.

Unsustainable land management practices that are degrading soils include: continuous cropping with reductions in fallow and rotations, repetitive tillage and soil nutrient mining; overstocking, overgrazing and burning of rangelands; and the overexploitation or clearance of wooded and forest lands. There is a need for identifying and promoting appropriate production systems and management practices that simultaneously reverse or minimize degradation, conserve above- and below-ground biodiversity, sequester carbon, reduce GHG emissions and at the same time ensure sustained agricultural production.

Controlling soil erosion: Soil erosion is caused by weather, wind, water and human induced activities. Studies have identified tillage-induced soil erosion as the major cause of the severe soil carbon loss and soil translocation on upland landscapes (Lobb et al., 1995; Lobb and Lindstrom, 1999; Reicosky et al., 2005). Runoff and resulting soil erosion can be substantially reduced through the adoption of minimum tillage or zero-tillage techniques combined with optimizing soil cover through cover crops, crop residues and mulch. On steeper slopes, soil erosion can also be reduced by planting cross-slope vegetation; using soil and water conservation structures, such as terraces, earthen bunds and tied ridges to optimize water capture and infiltration; and creating grassed waterways to convey excess water safely off the slopes. Measures to reduce erosion by wind include optimizing vegetation cover with drought-resistant species, using rotational grazing to sustain rangeland vegetation quality, planting windbreaks perpendicular to the prevailing winds and developing shelter belts.

Improving water storage: Water storage in the soil depends on many factors, including rainfall, soil depth, pore space, soil texture and soil structure. Soil management can influence rainwater infiltration and the capacity of the soil to reduce soil water evaporation and store water in the soil. Groundcover management can have highly beneficial effects on soil surface conditions, soil organic matter content, soil structure, porosity, aeration and bulk density. Improvements in these properties influence infiltration rates, water storage potential and water availability to plants.

Improving soil structure with organic matter: Usually soils are compacted due to repetitive hoeing or ploughing and soil compaction by trafficking of machinery. In grazing lands, soil is compacted by the trampling of livestock or wildlife. Soil compaction reduces airspaces in the soil and decreases the penetration of plant roots. Compacted soils and shallow soils are seriously affected by dry spells that limit root growth and the plant's access to moisture and nutrients. Sub-soiling, or ripping, is soil preparation treatment done with tined implements to break up hardpans without turning the soil upside down (FAO, 2012c). Prevention measure against soil compaction is minimum tillage in combination with a plant or litter cover which provides organic matter that enhances the activity of soil fauna (e.g. earthworms and termites). The burrowing of these soil organisms breaks up compacted layers and incorporates soil organic matter from the surface into the soil. Specific cover crops with strong and deep roots such as radish or pigeon peas can be used to penetrate and break up compacted soils layers. Conservation agriculture practices will allow soil organic matter to build up and increase the soil's resilience to climate change.

Managing soil organic matter for soil carbon sequestration: Elimination of perennial groundcover, repetitive cultivation or continuous grazing, bare fallows, removal of crop residues and grassland burning cause loss of soil organic matter and water holding capacity of soil. Monoculture of cash crops and the excess use of external inputs are adopted to achieve the highest possible yields and returns with minimal labour. Production of energy-intensive mineral fertilizers and pesticides and use of fossil fuel in farm machinery are major sources of GHG emissions. Further, when incorrectly applied, these inputs leach into water resources and the resulting water contamination has serious deleterious effects on ecosystems and human health. Diversified cropping, crop rotations and improved techniques for the management of fertilizer, seeds and pesticides can make the application of inputs more efficient and can reduce GHG emissions due to reduced use of inputs. By improving soil structure and increasing soil biodiversity, no-till cultivation and the control of soil compaction will also reduce GHG emissions, which result mainly from anaerobic soil conditions. There is common consensus that zero tillage and conservation agriculture systems will considerably reduce nitrate leaching from soil (Macdonald et al., 1989).

(ii) Water management for sustainable agriculture

Climate change accelerates crop evapotranspiration, changing rainfall patterns, and leads to greater variations in groundwater recharge including weather aberrations. The adverse effects of unprecedented droughts and floods are leading to groundwater depletion and severely affecting agricultural water management. Water management and automated irrigation systems can help accelerating the irrigation operations while ensuring substantial water conservation. A good farm management system can efficiently help water conservation and ensure a climate-resilient future of the farms.

Adaptation at farm level: Farmers prefer more efficient irrigation technologies that reduce evaporation losses. These technologies can be combined with deficit irrigation approaches to maximize productivity per volume of water applied in irrigation rather than per area of land. Crop selection and changes in crop calendars will help farmers adapt to new temperatures and rainfall patterns. The crops and varieties which are tolerant to dry conditions are preferred as they perform better than crops and varieties sensitive to dry conditions. Integrated farming system involving agriculture, horticulture, tree farming, livestock or fisheries reduces the risk of climate change.

Adaptation at irrigation level: Modern irrigation system, automated irrigation and using irrigation water on the basis of tensiometer etc. can economise irrigation water and increase the water use efficiency. Intermediate storage of water and conjunctive use of surface and ground water can economise use of irrigation water and build resilience and reliability of water supply. Water pricing on volumetric basis is often advocated as demand management tools for promoting efficient water use and reducing water wastage.

Adaptation at watershed, river basin and national levels: Flood management plans would require up-gradation of infrastructure with non-structural, information-rich approaches that can better mitigate the impact of floods through a combination of land planning, early warning system and crop insurance schemes. Similarly, there will be a need to shift from drought emergency response to drought management plans that include prevention, preparedness, relief and rehabilitation and long-term measures to mitigate the impacts of droughts (FAO and NDMC, 2008). Improved weather forecasting and hydrological monitoring will become a critical element of modern adaptation strategies (Faurès et al., 2010).

Water management for climate change mitigation: Usually irrigated agriculture uses more agro-chemicals for higher crop yields than rainfed agriculture. In absence of surface irrigation possibilities, ground water exploitation is made to increase the area under irrigation. The use of groundwater is expanding both in absolute and relative terms (Siebert et al., 2010), and this is increasing the use of fossil fuels and raising the energy costs of water supply. In India, it is responsible for 4–6 per cent of the national total emissions (Shah, 2009). Agricultural methane (CH_4) emissions account for more than 50 per cent of CH_4 emissions from human activities. Flooded rice cultivation releases methane in greater amount. Emissions during the growing season can be reduced by using various water management practices, such as cultivating aerobic rice and adopting alternate wetting and drying (AWD) method, where possible. Avoiding water saturation when rice is not grown and shortening the duration of continuous flooding during the rice growing season are effective options for mitigating CH_4 emissions from rice fields. System of Rice Intensification (SRI) is a practice that aims to increase the productivity of irrigated rice with resource conservation. In this system, younger seedlings are transplanted at square pattern under puddled condition and the field is kept saturated without submergence to allow more number of effective tillers. Weeds are incorporated into soil by mechanical weeding by a specialised weeder, which provides nutrition. As SRI reduces the amount of flooding of irrigated rice, it reduces CH_4 emissions and saves irrigation water. It may possibly reduce nitrous oxide emissions (HLPE, 2012).

Management of irrigated pastures, grazing management and use of feed additives can reduce emission of methane by ruminants.

8.2.2 Adaptation Strategies

Adaptation strategy is very important to reduce the severity of climate change impacts on agriculture. Various approaches can be taken up to adopt these strategies.

Adaptation through transgenic approaches: Transgenic approach involves manipulation of genetic material and the fusion of cells beyond normal breeding barriers. Genetically modified organisms (GMO) developed by genetic engineering can resist stressed conditions and provide resistance against certain pests. Identification and transfer of stress-tolerance gene into plant provide improved productivity and adaptation to abiotic stresses. Transgenic plant can be used as sources of new cultivars for abiotic stress resistance under weather variability.

Heat resistance: Plants have various mechanisms to cope with high temperatures. Such mechanisms may be maintaining membrane stability, or by ion transporters, proteins, osmo-protectants, antioxidants and other factors. Crop cultivars having such mechanisms have better tolerance to heat stress. Up-regulation of different genes helps the plant withstand the stress conditions and also adaptation to stress. New varieties tolerant to heat and extreme weather are now available for specific locations.

Drought tolerance: Several drought-tolerant transgenic plants have been developed by the breeders by genetic engineering. Stress-signalling networks in drought responses are made of intercellular communication systems, intracellular signalling systems and transcriptional-regulatory complexes. Some important proteins like protein kinases, receptor-like kinases, transcription factors, some enzymes related to osmo-protectant or plant hormone synthesis, and other functional and regulatory proteins play an important role in plant defence during drought stress. These crops and cultivars with drought resistance abilities are better suited to the areas frequently affected by drought.

Salinity tolerance: Climate change has caused accumulation of salts due to rise in temperature and soil water. The damaging effects of salt accumulation in agricultural soils have severely affected agricultural productivity in many areas. This problem can be tackled by using salt-tolerant varieties of crops and crop management practices. Transgenic plants accumulate more K^+, Ca^{2+} and Mg^{2+} and less Na^+ in their shoots compared with non-transformed controls. Saline tolerant crops and varieties are better suited to areas affected by salinity.

Developing climate-ready crops: Several drought tolerant varieties have been released and notified in India, which showed a consistently good performance under transplanted low-land conditions and rain-fed direct-seeded situations.

Crop varieties like Sahbhagi dhan, IR64-Drought-1, are climate-ready rice varieties. Singh *et al.* 2014 reported 108% higher yields with Sahbhagi Dhan compared with popular local varieties under water-scarcity situation (Haefelea SM, *et al.* 2016). Sahbhagi Dhan needed one to two irrigations only to yield 4.7 t ha^{-1}. Likewise, Sahbhagi dhan, IR64-Drought-1, are climate-ready rice varieties (Prasad *et al.* 2014).

Crop diversification: Instead of monoculture, crop diversification proves better adapted to weather variability. Mixed cropping and inter cropping systems are found to be more stress tolerant than sole crops. Intercropping of soybean + pigeon pea (4:2), pearl millet + pigeon pea (3:3), pigeon pea + green gram (1:2) and cotton + green gram (1:1) are more economic than sole cropping (Prasad *et al* , 2014). Due to growing income and urbanization, demand for high-value food products, such as fruits, vegetables, dairy, meat, eggs and fish, is increasing while demand for traditional cereals is reducing. Diversification of rice-wheat system towards high-value commodities will increase income and result in reduced water, fertilizer and other resource uses. Nitrogen-fixing pulse crops have a lower carbon and water footprint as compared to other crops. These crops are well adapted under diversified cropping system.

Alteration in land-use pattern: Change in location of crop and livestock, adjustment in cropping pattern, planting time and methods, fertilizer and pesticide use pattern, and other management practices help reduce the risk of climate change. Alternate land use systems such as tree farming, silvi-pastoral system, agroforestry etc. are recommended for rain-fed situations which can provide remunerative production under weather aberration.

Adjustment of planting time: The sowing and harvesting time can be manipulated as per weather condition of the location to reduce the impact of weather variability. Adjustment of planting time may reduce the impact of increased temperature on spikelet sterility and flowering.

Conservation agriculture: Conservation agriculture (CA) enhances natural biological process above and below the ground. It is characterized by three linked principles, namely (i) minimum mechanical soil disturbance, (ii) permanent organic soil cover by mulching, and (iii) diversification of crop species grown in sequences/association for risk minimization. Conservation agriculture-based resource conservation technologies (RCTs) under precision laser land levelling are helpful to improve water productivity (Jat *et al.* 2009).

Increasing resource use efficiency: Resource conserving practices include minimum tillage/zero-tillage, use of neem-coated urea, crop residue management, nutrient management, irrigation management etc. These resource conserving technologies (RCT) improve the resource use efficiency and conserve natural resources (water, soil, biodiversity) and reduce the production cost and environmental pollution. Neem oil-coated urea was found to be effective in reducing N_2O emission from zero-tillage wheat treatment (17.8–20.5%) leading to lower global warming potential as compared to conventional tillage wheat. Application of rice residue in zero-tillage wheat treatment also reduced N_2O emission (11–12.8%) (Gupta *et al.*, 2016).

Integrated nutrient management (INM): INM practice integrates application of plant nutrients through organic and inorganic sources, use of well-decomposed crop residues, green manures, recyclable waste, compost including vermi-compost, using legumes in cropping systems, use of bio-fertilizers and other locally available nutrient sources for sustaining soil health and amelioration of environment as well as enhancing crop productivity on long-term basis. Based on the study, it can be recommended that the substitution of 50% of the recommended N with organic source increases crop yields and soil carbon in semi-arid rain-fed systems of India (Prasad *et al.* 2016). INM can reduce the production cost, reduce pest attack and provide climate resilience.

Site specific nutrient management (SSNM): SSNM provides principles and tools for supplying nutrients as and when needed for plant to achieve high yields while optimizing the use of nutrients from indigenous sources. Nutrients should be applied at right time, right quantity and right method. Leaf Colour Chart (LCC) and GreenSeeker are used for assessment of nutrient application time in plants.

Relocation of crops: Climatic variability such as increased temperature, CO_2 level, drought and floods would affect the production of crops. There are sensitive crops which have to be identified and relocated in suitable areas.

Harnessing indigenous technical knowledge of farmers: Indigenous technologies (ITKs) are available in rural areas which provide adaptation to climate variability. These technologies may be validated and fine-tuned to overcome climate vulnerabilities. Such ecological-based ITKs can be used as adaptive measures.

Integrated Farming System (IFS): Mono-cropping in flood- and drought-prone area is risky practices for farmers. In rain-fed areas, the single enterprise may be exposed to climate risk and lead to crop failure. Integrated farming system (IFS) is defined as the integration of different interrelated, interacting and interdependent farm enterprises which are suited to agro-climatic condition socioeconomic situation of the farmers. IFS can minimise the risk in natural calamities and the diversified enterprises in the system can generate more income and employment in the year. Integrated fish-duck farming and rice-fish-poultry farming have been developed for small and marginal farmers (Prasad *et a*l. 2014).

Integrated Pest Management (IPM)

The dynamics of pest population and crop diseases are changing rapidly with the extreme changes in the climate. Many climatic factors such as variability in rainfall and changes in temperature would affect the incidence of pest, disease and host susceptibility of major crops, because climate is continuously changing which will potentially influence the pest/weed-host relationship by affecting the pest/weed population, the host population and the pest/weed-host interactions. The crops pests are controlled by natural defenders under ideal field situations. The farmers very often use chemicals for controlling crop pests without realising the important role of natural defenders which become counterproductive. Indiscriminate use of toxic chemical leads to resurgence of crop pests and such pests develop a resistance to such chemicals. IPM provides a lasting solution to pest management. It uses resistant/tolerant crop varieties, cultural control, physical and mechanical control, biological control and need-based chemical control when the pest population exceeds the economic threshold level (ETL). Many effective bio-agents are now available from laboratories against key pests and many improved traps like pheromone traps, yellow traps and sticky traps along with bio-pesticides are available for IPM.

Early warning system: Early warning system based on weather forecasting will guide the farmers to take up required measures against weather adversities. Information and Communication Technologies could greatly help the administrators and researchers in developing contingency crop plans and also reduce the risk of aberrant weather. Basing the local weather data early warning is given to farmers along with crop advisory to reduce the impact of climate change.

Weather based crop insurance: Loss in crop production and farm income due to unexpected weather hazards is beyond the carrying capacity of resource poor farmers of the country. The State and National governments on their part are coming to the rescue of farming community by implementing various relief measures like reduction or suspension of land revenue taxes, loan waiving, relief from calamity relief fund etc. In India, since 1972 agricultural insurance has evolved as an adaptation strategy to mitigate / manage weather related risks in agriculture. However, experiences with traditional agricultural insurance schemes have shown that this approach is not often suitable in developing countries. In recent years, weather index insurance contracts in agriculture have emerged as an alternative to traditional agricultural (or crop) insurance.

8.2.2 Mitigation Strategies

Mitigation practices help reduce the emission of greenhouse gases from agriculture system. A change in the system of cultivation, cultural practices, nutrient management and water management helps reduce the emission of greenhouse gases.

Methane: Methane is about 25 times more effective as a heat-trapping gas than CO_2. In agriculture, methane gas is released from wetland rice cultivation, organic decay, enteric fermentation of cattle dung etc. Production, oxidation and transport of CH_4 from paddy field can be done by alteration in water management, particularly promoting mid-season aeration, by short-term drainage, direct-seeded rice (DSR) cultivation. Reduction in CH_4 emissions increased further when DSR was combined with mid-season drainage or intermittent irrigation, compared with flooded transplanted rice. The reason for low CH_4 emissions from DSR is aerobic conditions, especially during the early growth stage. The other methods for reducing methane are improved organic matter management by promoting aerobic degradation through incorporating or composting it into soil during off-season drained period, cultivation of rice varieties with high root oxidative activity and high harvest index, and feed management in livestock by using improved feed additives, substituting low digestible feed with high digestible feed and grazing which can reduce methane emission in livestock farming.

Nitrous oxide: Nitrous oxide (N_2O) is more effective than CO_2 as greenhouse gas. The major sources of N_2O are forests, grasslands, oceans, soils, nitrogenous fertilizers and burning of biomass and fossil fuels. Site-specific nutrient management (SSNM) in crop fields, integrated nutrient management (INM), use of slow-release nitrogen fertilizers and nitrification inhibitors, and placement of fertilizer in reduced zone helps to reduce emission of N_2O.

Carbon dioxide: The main sources of carbon dioxide emission are decay of organic matter, forest fires and eruption of volcanoes, burning of fossil fuels, deforestation and land-use changes. Such emissions could be controlled by carbon sequestration, soil management practices and planting trees. Carbon sequestration is one of the best strategies to mitigate CO_2 emission, which is the removal of atmospheric CO_2 into long-lived stable form.

Carbon sequestration: Trees, plants and crops absorb CO_2 naturally through photosynthesis and stored as carbon in biomass in tree trunks, branches, foliage and roots and soils. Trees function as carbon sinks and store large amount of carbon in their vegetation and root system.

Soil management practices: Soil management practices such as reduced tillage and manuring, residue incorporation and mulching can improve soil biodiversity, micro-aggregation and improve soil health and water holding capacity. Conservation tillage reduces soil disturbance and helps mitigate the release of soil carbon to the atmosphere.

Improved cropping and organic systems: Organic systems of production increase soil organic matter levels through the use of composted animal manures and cover crops. Addition of sufficient organic manure can dispense with application of chemical fertilisers. Reduced carbon dioxide emissions and increased soil carbon can be achieved by direct seeding, field windbreaks, rotational grazing, perennial forage crops, reduced summer fallow and proper straw management. Using higher-yielding crops or varieties and maximizing yield potential can also increase soil carbon. Conservation practices such as conservation tillage, integrated and site specific nutrient management, crop rotation and diversification, management of animal manure and feed management in livestock can reduce emission of GHG.

Land restoration and land use changes: Land restoration and land use changes that encourage the conservation and improvement of soil, water and air quality can reduce greenhouse gas emissions. Modifications to grazing practices, such as implementing sustainable stocking rates, rotational grazing and seasonal use of pasture land, can lead to greenhouse gas reductions. Conversion of marginal lands and fallow lands to tree plantation can enhance carbon storage.

Livestock management: About one-fifth of global CH_4 emissions are from enteric fermentation in ruminant animals. Animal breeding and genetic selection for feed efficiency can reduce CH_4 emissions. Changing the composition of animal diet, use of feed additives, feed management, grazing management and cultivation of better forage plants can reduce the CH_4 emission from enteric fermentation.

Mitigation through transgenic crops : As indicated by researchers, there is a need to give priorities to develop new varieties of crops cultivars that can reduce GHG emissions in farming systems (Barfoot *et al.,* 2014). In this regard, Genetically Modified (GM) crops have been contributing to lower GHG emissions through reducing fuel use, due both to less pesticide applications and increasing the area under conservation agriculture. There are various genetic engineering activities which improve nitrogen use efficiency in crops. Engineering nitrogen fixation for self-fertilizing crops can also help reduce the use of synthetic fertilizer.

Irrigation and water management: Improvements in water use efficiency coupled with a reduction in operating hours, drip irrigation technologies can significantly reduce the amount of water and nitrogen applied to the cropping system.

Nitrogen use efficiency: Precision farming and use of site-specific nutrient application practices can reduce GHG emission. Other strategies include the use of cover crops and manures (both green and animal), rotation with nitrogen-fixing crops, application of compost and bio-fertilisers with better fertiliser application methods required to increase nitrogen use efficiency.

Methane capture: Large emissions of methane and nitrous oxide are attributable to livestock waste treatment, especially in dairies. Agriculture methane collection and combustion systems include covered lagoons and complete mix and plug flow digesters. Bio-gas plants have become popular in many areas for production of methane gas for fuel and home lighting. Additional benefits include reducing odours from livestock manure and reducing labour costs associated with manure removal.

Biofuels: Bio-fuels in form of bio-diesel or ethanol are carbon neutral. Bio-fuels can be produced from crop residues and sugarcane trash. Renewable energy sources such as wind and solar also present significant opportunities for the agriculture sector to reduce greenhouse gas emissions.

References

Barfoot P, Brookes G. Key global environmental impacts of genetically modified (GM) crop use 1996-2012. GM Crops and Food: Biotechnology in Agriculture and the Food Chain. 2014;5(2):149-16

FAO 2019. Climate-smart agriculture and the Sustainable Development Goals: Mapping interlinkages, synergies and trade-offs and guidelines for integrated implementation. Rome. file:///C:/Users/User/ Downloads/CA6043EN.pdf.

FAO, 2013.http://www.fao.org/do crep/018/i3325e/i3325e00.htm. Climate Smart Agriculture Source Book. Retrieved on April,2017.

FAO. 2014. Transitioning towards climate-smart agriculture in Kenya. P[policy brief]. Food and Agriculture Organization of the United Nations, Rome. (available atRetrieved from www.fao.org/3/ai4259e.pdf.)

FAO. 2012. LADA local level manual for the assessment of land degradation and sustainable land management. Rome. (available at www.fao.org/nr/lada)

FAO. 2007. Modernising irrigation management – the MASSCOTE approach. FAO irrigation and drainage paper No. 63. Rome

FAO & National Drought Mitigation Center (NDMC). 2008. The Near East drought planning manual: guidelines for drought mitigation and preparedness planning. Food and Agriculture Organization of the United Nations Regional Office for the Near East, Cairo and University of Nebraska-Lincoln, National Drought Mitigation Centre

Faurès, J.M., Bernardi, M. & Gommes, R. 2010. There is no such thing as an average: how farmers manage uncertainty related to climate and other factors. International Journal of Water Resources Development, 26: 4, 523—542

Food and Agriculture Organization of the United Nations. 2010. "Climate-Smart" Agriculture, Policies, Practices and Financing for Food Security, Adaptation and Mitigation.

Gupta DK, Bhatia A, Kumar A, Das TK, Jain N, Tomer R, Malyan SK, Fagodiya RK, Dubey R, Pathak H. Mitigation of greenhouse gas emission from rice–wheat system of the Indo-Gangetic plains: Through tillage, irrigation and fertilizer management. Agriculture, Ecosystems and Environment. 2016;230:1-9

Haefelea SM, Kato Y, Singh S. Climate ready rice: Augmenting drought tolerance with best management practices. Field Crops Research. 2016;190:60-69

High Level Panel of Experts on Food Security and Nutrition (HLPE). 2012. Climate change and food security. A report by the HLPE of the Committee on World Food Security, Rome

Jat M, Gathala MK, Ladha JK, Saharawat YS, Jat AS, Kumar Vipin, Sharma SK, Kumar V, Gupta R. Evaluation of precision land leveling and double zero-till systems in the rice–wheat rotation: Water use, productivity, profitability and soil physical properties. Soil & Tillage Research. 2009;105:112-121

Lobb, D., Kachanoski, R.G. & Miller, M.H. 1995. Tillage translocation and tillage erosion on shoulder slope landscape positions measured using 137Cs as a tracer. Canadian Journal of Soil Sci., 75: 211–2

Macdonald, A.J., Powlson, D.S., Poulton, P.R. & Jenkinson, D.S. 1989. Unused fertiliser nitrogen in arable soils-its contribution to nitrate leaching. Journal of the Science of Food and Agriculture, 46: 407–419

Prasad YG, Maheswari M, Dixit S, Srinivasarao CH, Sikka AK, Venkateswarlu B, Sudhakar N, Prabhu Kumar S, Singh AK, Gogoi AK, Singh AK, Singh YV, Mishra A. Smart Practices and Technologies for Climate Resilient Agriculture. Hyderabad: Central Research Institute for Dryland Agriculture (ICAR); 2014. p. 76

Prasad JVNS, Rao CS, Srinivas K, Jyothi CN, Venkateswarlu B, Ramachandrappa BK,Dhanapal GN, Ravichandra K, Mishra PK. Effect of ten years of reduced tillage and recycling of organic matter on crop yields, soil organic carbon and its fractions in Alfisols of semi arid tropics of southern India. Soil & Tillage Research. 2016;156:131- 139

Reicosky, D.C., Lindstrom, M.J., Schumacher, T.E., Lobb, D. & Malo, D.D. 2005. Tillage induced CO2 loss across an eroded landscape. Soil and Tillage Research, 81 (2): 183–194

Scherr, S.J. & Sthapit, S. 2009b. Sustainable land management in Africa. TerrAfrica Climate Briefs No. 1 & 2

Siebert, S., Burke, J., Faures, J. M., Frenken, K., Hoogeveen, J., Döll, P.& Portmann, F. T. 2010. Groundwater use for irrigation – a global inventory. Hydrology and Earth System Sciences, 14: 1863–1880

Singh A, Yadav A, Matthew K, Varrieur V, Singh S. Success of IRRI stress tolerant rice varieties in Bihar, India-Experience of IRRAS farmers. In: Poster Presented at Fourth International Rice Conference; 27 October to 01 November 2014; Bangkok, Thailand. 2014

9

Climate Smart Agriculture Practices

9.1 Climate-smart Agriculture Practices and Indicators

Considering the three pillars of climate-smart agriculture such as productivity, adaptation and mitigation, different technologies and practices have been evolved.

The CSA practices are grouped under soil and land management, water management, crop management, livestock management, forestry, fishery & aquaculture and energy management. The World Bank Group (2016) has suggested broad indicators of CSA practices (Table 7.1; Chapter 7)

9.2. Climate-smart Agriculture Practices

The CSA practices can be grouped into water management, soil management, crop management, livestock& fisheries management and forest management

9.2.1 Soil and Land Management

These practices can adapt to the adverse effects of climate change and reduce GHG emissions from agriculture and build resilience in farming systems.

Preventing land conversion and protecting vulnerable lands to high soil organic matter losses: Cultivated lands vulnerable to loss of soil organic matters at a greater rate need protection. Soil carbon stocks in semi-arid environments can decrease by 30 per cent in less than five years when native vegetation or pastures are converted to cropland (Zach et al., 2006; Noellemeyer et al., 2008). Cultivation of tropical forest soils causes losses of more than 60 per cent of original soil organic matter (SOM) content stocks in just a few years (Brown and Lugo, 1990). Although drylands have lower mitigation potential per hectare than humid lands, their overall contribution could be highly significant since dry and sub-humid lands cover 47 per cent of the Earth's land surface. Moreover, many dryland regions affected by land degradation are sub-optimally managed could rapidly respond to improvements in management (Farage et al., 2007).

Assessment of land resources is required to understand the extent of land degradation and identify hot spots in terms of land degradation. Suitable land use planning and policy support or incentives are required to reduce land conversion and promote the adoption of sustainable practices, with particular attention to lands that are more vulnerable to climate change.

Preventing and mitigating land degradation: Degraded soils are vulnerable to serious losses of SOM and soil biodiversity, greater soil compaction and increased rates of soil erosion and landslides. Land degradations are physical degradation by water and wind, crusting, sealing and waterlogging; biological degradation due to the depletion of organic matter depletion and loss of soil flora and fauna; and chemical degradation by acidification, nutrient depletion, pollution from excessive use of pesticides and fertilizers or human and industrial waste. The major reasons for soil degradation are continuous cropping without rotation and fallowing, repeated tillage, over grazing, burning and soil nutrient mining. Sustainable management practices that would reverse or minimize degradation, conserve bio-diversity, sequester carbon, reduce GHG emissions and ensure sustainable productivity would be required. Such practices are crop rotation, fallowing, organic recycling and minimum tillage.

Controlling soil erosion: The practices to decrease the soil erodibility and runoff are usually adopted for erosion control and soil conservation. Tillage induced soil erosion is a major cause of soil carbon depletion. Soil erosion caused by rain water, wind and other factors can be prevented by many conservation practices. Avoiding soil burning, optimizing vegetative cover with adapted species, using rotational grazing to sustain the quality of rangeland vegetation and enhancing the roughness of the soil surface with tied ridges, contour farming and planting wind breaks are important practices (Fig.9.1).

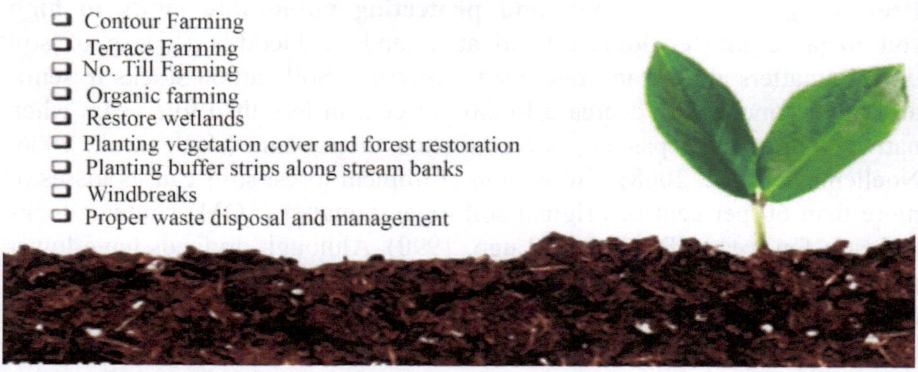

Fig. 9.1: Soil conservation practices

Water erosion can be controlled by the adoption of minimum to no tillage techniques combined with optimizing soil cover (cover crops, residues, mulch). Soil conservation measures like cross-slope vegetation, terrace cultivation, contour bunding, contour cultivation, tied ridges, grassy water ways, trenches, field bunding, half-moon terraces can reduce soil erosion caused by run-off water. Measures to reduce wind erosion include optimizing vegetation cover with drought-resistant species, using rotational grazing to sustain rangeland vegetation quality, shelter belts and planting windbreaks perpendicular to the prevailing winds.

Improving water storage: Water storage in soil is governed by factors like rainfall, soil depth, soil texture and soil structure. Since surface evaporation, evapotranspiration from plants and ground water movement influence loss of water from soil, soil management can influence rainwater infiltration and the capacity of the soil to reduce soil water evaporation and increase water storage in the soil. Ground cover management can have beneficial effects on soil surface condition, soil organic matter content, soil structure, porosity, aeration and bulk density of soil. Improvements in these properties influence infiltration rates, water storage potential and water availability to plants. The relatively available soil moisture to plants usually is the difference between field capacity and permanent wilting point. Soil management can reduce rates of erosion, the dispersion of soil particles and the risks of waterlogging and salinity in drylands. In a soil matrix, stable forms of organic carbon, such as humus, can hold larger amount of water. The soil management practices that are required for improving water storage in soil are minimum soil disturbance, mulching, precision farming, and micro-irrigation (sprinkler or drip).

Improving soil structure with organic matter: Soil organisms break down the organic matter to use it as food and to release excess nutrients into the soil for use by plants. Soil organic matter primarily contains organic carbon (on average 58 per cent), but also nutrients that are essential for plant growth, and some inorganic carbon. It influences the physical, chemical and biological properties of soil. The factors responsible for soil organic matter decomposition are presence of soil organisms, temperature, moisture, and physical and chemical properties of soil along with land use pattern and natural resource management practices.

Minimum tillage: The basic function of soil tillage is to control weeds and provide a suitable seedbed for rapid germination and development. Tillage can also be defined as the mechanical manipulation of soil with certain implements or tools to provide a suitable environment for seed germination, root growth and weed control. Soil compaction is caused by the passing of heavy machinery through the fields for tillage and repeated tillage. In grazing lands soil is compacted by the trampling of livestock or wildlife. Soil compaction reduces porosity and decreases the penetration of plant roots resulting restricted growth and movement of lateral roots. There is often hard pan development below the soil surface which requires sub-soiling. Tillage disturbance in soil causes reduction in carbon stabilisation. Further more, tillage of the soil stimulates microbial decomposition of soil organic matter, which results in emissions of CO_2 to the atmosphere.

The management practices to prevent soil compaction and hard pan development are minimum tillage, plant or litter cover to provide organic matter that enhances the activity of soil fauna (e.g. earthworms and termites). The burrowing of earthworms can break up compacted layers and incorporating SOM from the surface into the soil. Practices such as conservation agriculture (that combine minimized soil disturbance with increased soil cover and crop diversification) will allow SOM to build up and increase the soil's resilience to climate change. Biological-tillage activity of macro-fauna (such as earthworms) that burrow and make channels for air and water breaks down organic matter in the soil.

Minimum tillage or conservation tillage improves the biological, chemical and physical properties of soil. The minimum tillage concept reduces time, labour and machine operations as well as conserving moisture and reducing erosion. In the last decades, advancements in weed control methods and farm machinery have allowed many crops to be grown with minimum tillage. There are three types of reduced tillage: conservation, zero and minimum tillage. Conservation tillage increases the ability of soil to store or sequester carbon while enriching the soil and improving soil water infiltration. This reduces erosion and nitrate runoff, improves the stabilization of soil surface to wind erosion, reduces leaching of nutrients due to greater amounts of soil organic matter to provide binding sites, decreases evaporation and increases soil moisture retention, which can increase yields in drought years. Conservation tillage also reduces the cost of fossil fuel and the associated carbon emissions to the atmosphere.

Zero tillage (also called no-till farming or direct drilling) is a way of growing crops or pasture from year to year without disturbing the soil through tillage. No-till is an agricultural technique which increases the amount of water that infiltrates the soil and increases organic matter retention and cycling of nutrients in the soil. In zero tillage, crop residues are spread on the soil surface, weeds are controlled by cover cropping or use of non-pollutant desiccant herbicide as pre-plant application, and use of specialised planter for planting of the seeds with minimum soil disturbance. Crop rotation is fundamental to zero tillage because it helps minimize weed, insect and disease populations that increase when the same crop is grown year after year on the same ground.

Managing soil organic matter for soil carbon sequestration: The rate of organic matter decomposition of soil depends primarily on the soil biota, temperature, moisture and chemical and physical composition of soil. It is also affected by the previous land use and natural resource management practices such as tillage. Loss of soil organic carbon and water holding capacity of soil is associated with practices such as the elimination of perennial groundcover, repetitive cultivation or continuous grazing, bare fallows, removal of crop residues and stubble burning. The monoculture of cash crops uses energy-intensive fertilisers and pesticides as a result of which there is GHG emissions. Excess and improper use of such chemicals leach into water resources and the resulting water contamination has serious deleterious effects on ecosystems and human health. The management practices such as crop rotations and improved techniques for the management of fertilizer, seeds and pesticides can increase resource use efficiency and reduce emissions of GHG gas. Improvement of soil structure and soil biodiversity, no-till cultivation and the control of soil compaction will reduce GHG emissions.

Nutrient management: In conventional agriculture organic manures are supplemented by synthetic fertilisers to supply macro-nutrients and micro-nutrients as per crop's requirement. The excessive nutrient mining of the soil has to be managed by soil fertility/nutrient management systems that take the strong crop-livestock interactions into consideration (FAO, 2013). The options are inorganic fertilisers, grain legumes, animal manures, integrated nutrient management and agroforestry options appropriate to the area and people.

Fertiliser Tree Systems (FTS) involve the planting or regeneration of fast-growing nitrogen-fixing trees or woody shrubs to produce enough biomass and to adapt to local climate. Vermi-compost is used as a better source of organic fertiliser. Many crop residues and animal manures are of low quality, as they fall below the critical nitrogen content of 1.8 to 2.0% and immobilise nitrogen temporarily, thereby exacerbating the nutrient deficiency. A strategy used worldwide in intensified cropping systems is to combine inputs of organic matter (mulch, compost, crop residues, green manure) with fertilisers to address or prevent macro- and micro-nutrient deficiencies (FAO, 2013).

Through the activity of soil micro-organisms, organic nitrogen in soil is mineralized as ammonia (NH_3) and ammonium ions (NH_4 -) and then transformed (nitrified) into nitrates (Jiang and Bakken, 1999). Since nitrate ions are not attracted to clay and soil organic matter particles, they move freely with soil water and leach through soil drainage. In absence of oxygen, denitrifying organisms will reduce nitrate to nitrous oxide. Management practices that can reduce such nutrient loss are ensuring timing and type of nitrogenous fertiliser, adding nitrification inhibitors, using slow release fertilisers like neem coated or sulphur coated, and adding nitrification inhibitors containing ammonium to fertilizer. Use of slow-release fertiliser, nitrification inhibitors, higher use efficiency, multiple applications and application at lower rates can reduce losses of reactive nitrogen due to leaching and runoff, as well as losses from volatilisation and mineralisation.

There is common consensus that zero tillage and conservation agriculture systems will considerably reduce nitrate leaching (Macdonald et al., 1989). Cover crops take up the nitrogen and reduce its loss from the soil. At the same time, unused mineralized nitrogen remains distributed within smaller pores and are not washed out of the soil (Bergström, 1995; Davies et al., 1996; Gors et al., 1993).

Biochar, a stable and carbon-rich form of charcoal is produced by pyrolysis, where biomass is heated with little or no oxygen (Sparkes and Stoutjesdijk, 2011). Possible biomass sources for *biochar* include: milling residues (e.g. rice husks, sugar cane bagasse); crop and logging residues; biofuel crops; municipal wastes; and animal manure. Application of *biochar* is soil adds nutrients and retains nutrients in soil from other sources, as a result of which the growth and yield of crop can be increased.

Contour farming and terrace farming: Contour farming is practiced in hilly slopes along the contour lines with ridge and furrow system of planting to change the direction of runoff. This practice applies on sloping land where annual crops are grown. Orchards, vineyards and nut crops use this practice. Contour farming reduces sheet and rill erosion and reduces transport of sediment, other solids and the contaminants attached to them. Bench terraces are formed along the contour and across the slopes for planting in hill slopes.

Fig. 9.2: Contour farming

Mulching and residue management: Mulching and residue management can be defined as a technology whereby, at the time of crop emergence, at least 30% of the soil surface is covered by organic residues from the previous crop (Erenstein, 2002). Mulching improves infiltration, and the soil cover provides effective protection against rain splash erosion and surface runoff.

Instead of burning, the crop residues are required to be incorporated into fields for soil and water conservation, and building the soil organic matter content. Mulching offers opportunities to address soil fertility and weed management constraints and improves infiltration, provides effective protection against rain splash erosion and surface runoff.

Sustainable mechanization: Appropriate small farm machinery can carry out sustainable crop management practices which could increase production efficiency of farms. Small tractors used for small working hours and combined with conservation agriculture can minimise soil disturbance, reduce carbon dioxide emission and reduce soil erosion and land degradation. These smaller tractors including two-wheeler tractors use minimum amount of fuel. Equipment used in precision farming can improve efficiency in the use of inputs and reduce direct and indirect greenhouse gas emissions. Investments in sustainable mechanization can enable farmers to expand the range of their activities and diversify their livelihoods in ways that can reduce their vulnerability to climate change. Use of drones powered by battery can carry out efficient operations without emitting GHG. Solar powered pumps can be used for irrigation.

9.2.2 Water Management

Climate change will have an impact on every element in the water cycle (UN-Water, 2010).Agriculture production will be affected by increased evaporative demand of crops, changes in the amount of rainfall and variations in river runoff and groundwater recharge, which are basic sources of irrigation. The hydrological cycle is expected to accelerate as rising temperatures increase evaporation from land and sea (Turral et al., 2011). Rise in temperature may increase the rate of evapotranspiration, which may lead to depletion of soil moisture.

Options for adaptation to climate change will need to be applied at different scales: on fields and farms; in irrigation schemes (particularly in large schemes); in watersheds or aquifers; in river basins (including trans-boundary river basins); and at the national level.

Adaptation at field and farm level : Farmers will be inclined to adopt more efficient irrigation technologies that reduce evaporation losses. Deficit irrigation, crop selection and changes in crop calendars, use of crops or varieties with better resilience to dry spells, increased agricultural diversification, including better integration of trees, crops and livestock can provide resilience to climate change at farm level. Farmers will also need to adopt relevant measures to respond to increased frequency of floods and more intensive rainfalls. Both irrigation and drainage have to be integrated in irrigation systems.

Adaptation at irrigation scheme level: In the context of climate change, irrigation reduces the evapotranspiration and saves the crop from dry spells. Governments incur huge expenditure for development of irrigation sources to ensure an assured crop under climate change and provide scope for multiple cropping. In a climate-smart agriculture, farmers are motivated to use sprinklers and drips to increase water use efficiency of 70% to 90%.In lift irrigation renewable energy pumps (e.g., solar-powered pumps) are preferable as there is no consumption of fossil fuel and emission of greenhouse gases. Under surface irrigation, irrigation scheduling and management decisions such as supplementary, full and deficit irrigation can increase water use efficiency and stabilise crop productivity. In order to reduce conveyance losses in irrigation canals, rehabilitation of canals and lining can be done. Conjunctive use of surface and ground water along with rain-water harvesting will be beneficial in augmenting scarce water sources.

Irrigation modernization requires better water allocation mechanisms, the clear transmission of alerts about water scarcity to farmers, and the adaptation of both infrastructure and management for more flexible and reliable delivery of water (FAO, 2007). Intermediate water storage in ponds, combining ground water with surface water, pricing of water to reduce misuse of water etc. are some of the options. Management of water allocation, cropping pattern and water distribution by water users association (WUA) are simpler and more effective for inducing more productive water use in the irrigation system level.

In order to face of climate change, irrigation will be required to counter the predicted increase in evapotranspiration rates and decreases in rainfall in the future. However, there is need for change in the irrigation system like drip or subsurface irrigation that has higher field-level application efficiencies of 70% to 90%. Photovoltaic or solar-powered drip-irrigation systems combine the efficiency of drip irrigation with the reliability of a solar-powered water pump. In existing irrigation systems, conveyance loss should be reduced to increase water use efficiency.

Adaptation at watershed, river basin and national levels: Climate change adaptation at higher levels requires a combination of policy adjustments and investments in infrastructure and management. Adjustments in the storage capacity and management of dams and river protection works would be required at water basin level. Flood management plans will need approaches that can better mitigate the impact of floods through a combination of land planning, early warning and insurance schemes. Drought management plans should include prevention, preparedness, relief and rehabilitation measures.

Integrated water resource management (IWRM) on the basis of river basins is the most accepted strategy for river basin management as actions in one part of a basin will have repercussions for people further downstream.

Rainwater harvesting : Rain water harvesting involves collection and storage of rain water at surface or in sub-surface aquifer, before it is lost as surface run off. Rainwater conservation practices can be divided into two main categories: 1) *in-situ* and 2) *ex-situ* management. *In situ* conservation refers to harvesting rainwater where it falls in soil profile. Land and water conservation practices are adopted in the field itself for collection and storing of run-off rain water. *In situ* conservation of rainwater considers the factors like slope, soil type, vegetation, crops, rainfall pattern and tillage practices (Table.9.2).

Table 9.2: *In situ* water conservation practices

Soil type	Rainfall	In situ moisture conservation techniques
Red soil	Low	Dead furrow at 3-6 m intervals
	Medium	Sowing on flat beds and ridging afterwards
	High	Graded border strips
Black soil	Low	Contour cultivation
	Medium	Dead furrow at 3-6 m interval
	High	Graded open furrow at 10 m intervals across the slope

Location specific measures for *in situ* moisture conservation are conservation tillage, deep tillage, contour cultivation, dead furrows, strip cropping, contour farming, mulching, broad bed and furrow (BBF), vegetative barrier, ridge and furrow and tied ridging etc. These practices are taken up to protect the land from degradation and to encourage storage of water for crop use.

Fig. 9.3: Staggered contour trenches

Fig. 9.4: Bench terrace

Fig. 9.5: Vegetative barrier

Since a significant amount of water is lost through surface runoff, rain water harvesting and recycling are very important, especially in dryland areas. Rain water harvesting is done by *in-situ* harvesting, *ex-situ* harvesting and domestic harvesting. *In-situ* rain water harvesting is catching the rain where it falls. *Ex-situ* harvesting is external water harvesting of runoff water and storing it elsewhere for future use. In domestic water harvesting, water is collected from roof-top and courtyards.

In *in-situ* rain water harvesting different practices like off-season tillage, tied ridges, mulching, retention of crop residues, contour planting, ridging/ furrowing, terracing, vegetative barriers, stone lines and planting basins etc. are practiced. These technologies increases infiltration time and increase surface storage of run-off water.

External water harvesting can be categorised into micro-catchment and macro-catchment water harvesting. Micro-catchment water harvesting involves collecting runoff from a small drainage area and storing it in the root zone. Macro-catchment water harvesting takes place on a larger scale and involves capturing rock water or runoff from hill-slope catchments. Sub-surface water harvesting is made by constructing barrages and cut-off walls.

In *ex-situ* conservation practice rain water is collected and stored for different productive use such as irrigation, drinking water and sanitation etc. External water harvesting can be micro-catchment and macro-catchment water harvesting, and flood and groundwater harvesting. In micro-catchment water harvesting usual methods used are dugout ponds, diversion bunds, *khadins* and *haveli* system of rain water harvesting. In macro-catchment water harvesting is taken up on larger scale to capture run-off water from hill slopes. Sub-surface water harvesting is done by constructing barrages and cut-off walls. Barrages can be constructed from stone, concrete or compacted earth. In rural areas the techniques adopted for rain water harvesting are gully plugging, contour bunding, field bunding, percolation tanks, check dams, dug-wells, dams and sub-surface dykes.

Fig. 9.6: Dugout-cum embankment pond

Fig. 9.7: Dug-out pond with cement lining

Fig. 9.8: Check-dam

Fig. 9.9: Farm pond

Water management for climate change mitigation: Since irrigated agriculture uses greater amounts of inorganic fertilizer and other agrochemicals than most rain-fed systems, efforts are needed to reduce greenhouse gases through improved crop management practices. Ground water exploitation is increasing the use of fossil fuels and raising the energy costs of water supply. Renewable source of energy use is an option to reduce emission of GHG.

One-third of methane emissions from agriculture come from flooded rice production. Emissions during the growing season can be reduced by using various water management practices, such as cultivating aerobic rice and, where conditions allow, adoption of alternate wetting and drying system. System of rice intensification (SRI) reduces the amount of flooding of irrigated rice and it is also likely to reduce CH_4 emissions. It saves water and other resources. Better pasture management, combined with the use of feed additives that suppress CH_4 fermentation in ruminants, can reduce livestock CH_4 emissions substantially (Turral et al., 2011).

Water management in rice: Rice is a water guzzling crop and consumes maximum amount of water. The alternate wetting and drying method of irrigation of rice implies that rice fields are not kept continuously submerged but are intermittently dried during the rice growing. This method is very relevant for climate change adaptation and mitigation. Flooded soils such as those for irrigated rice produce methane, a greenhouse gas that plays a significant role in global climate change (Lindau et al., 1993 in van der Hoek et al., 2001). The research done by the International Rice Research Institute (IRRI) and national rice research institutes in China, India, Indonesia, the Philippines, and Thailand concluded that when rice fields are dried, oxygen becomes available in the root zone, and this reduces methane emissions. Producing more rice with less water from irrigated systems could provide opportunities to improve human health as some diseases like malaria can spread under submerged condition.

9.2.3 Climate-smart Crop Production

Several efficient crop management practices which address all the three aspects of CSA should be considered as "climate-smart" practices. The principles of climate-smart crop management are sustaining the crop yield by increasing input use efficiency and enhancing intrinsic system of climate resilience. The practices are use of stress tolerant varieties, change in cropping pattern, adoption of farming system approach and propagation of ecosystem management options. Integrated soil-crop-water management increases resilience to climate change.

Important crop management practices in the contest of CSA include direct seeding in combination with protective soil cover, crop diversification and crop rotation, the elimination of the burning of crop residues, integrated soil fertility management to increase the soil's nutrient retention capacity and the availability on nutrients to plants, the precise management of nitrogen and integrated pest management etc.

It has been predicted that billions of people living in developing countries will face shortages of water and food and greater risks to health and life due to climate change. Although some crops in some regions of the world may benefit, the overall impacts of climate change on agriculture are expected to be negative (IFPRI, 2009).

Localized extreme events and sudden pest and disease outbreaks are already causing greater unpredictability in production from season to season and year to year, and require rapid and adaptable management responses (FAO-PAR, 2011). Improving adaptation of the agricultural sector to the adverse effects of climate change will be imperative for protecting and improving the livelihoods of the poor and ensuring food security (FAO, 2012a). There are various options to make cropping resilient to climate change.

Sustainable crop production intensification: The Green Revolution was characterized by the planting of high-yielding crop varieties, with the associated chemical package and irrigation resulting cereal food production from 800 million tonnes to over 2.2 billion tonnes between 1961 and 2000 in India. While an estimated one billion people were saved from famine, intensive cropping have degraded fertile land, depleted groundwater, triggered an upsurge in pests, eroded biodiversity and polluted the air, water and soil. This has necessitated a second green revolution.

A productive agriculture that conserves and enhances natural resources through an ecosystem approach capitalizes on natural biological inputs and processes. It reduces the negative impacts on the environment and enhances natural capital and the flow of ecosystem services. The climate-smart crop production system requires practices that are based on improving efficiencies and managing biological processes.

The management practices include: (i) maintaining healthy soil; (ii) cultivating a wider range of species and varieties in sole cropping, rotations and sequences; (iii)using quality seeds and planting materials of well adapted, high-yielding varieties; (iv) adopting the integrated management of pests, diseases and weeds; (v) managing water efficiently, (vi) biodiversity conservation; (vii) agroforestry, and ((viii) integrated farming system. Both crop production and climate change adaptation and mitigation require a resilient ecosystem, which can be attained through approaches and practices that are based on the sustainable management of biodiversity and ecosystem services. In terms of technologies, productive and sustainable agricultural systems make the best of crop varieties and livestock breeds and their agro-ecological and agronomic management (Beddington et al., 2012).

Cropping system management: Monoculture with cash crops has become vulnerable to climate change. Mixed cropping, diversified cropping and inter cropping have become resilient to climate change.

(i) Intercropping: Intercropping is the practice of growing two or more crops on a piece of land. The crops should not be competitive to each other for physical space, nutrients, water and sunlight. Usually a shallow rooted crop is inter cropped with deep rooted crop. Inter cropping produces higher yield and land equivalent ratio (LER) than a sole crop. The yields in intercropping are more stable than sole crops. Some of the more significant types are mixed intercropping, row cropping and relay cropping. In mixed inter cropping, the component crops are mixed at a definite proportion and sown. Row cropping involves the component crops arranged in alternate rows. Variations include alley cropping, where crops are grown between rows of trees, and strip cropping, where multiple rows, or a strip, of one crop are alternated with multiple rows of another crop. Intercropping under scattered trees is the simplest form of agroforestry. The traditional intercropping systems consist of growing agricultural crops under scattered or systematically planted trees on farmlands. There are more diverse multi-storied home gardens in humid areas and less diverse, two-tiered canopy configurations (trees + crop) in drier areas.

Fig. 9.10: Intercropping system

(ii) Crop rotation: Crop rotation is a practice of growing different crops on the same land in a sequence in different seasons. Using some forms of crop rotation, farmers can reduce the need for artificial fertilizers as well as keep their fields under continuous production instead of letting them lie fallow. Crop rotation gives various benefits to the soil such as soil enrichment through nitrogen fixation by legumes.

Crop rotation invariably reduces the pest and disease attack and increases soil fertility. In the components of poly-culture, different crops can also reduce the effects of adverse weather for the individual farmer and, by requiring planting and harvesting at different times, allow more land to be farmed with the same amount of machinery and labour.

Fig. 9.11: Crop rotation

(iii) Crop production and adjustment of crop variety: A sound crop production system needs to adapt to climate change and become climate resilient. Farmers are using drought resistant or early maturing crop varieties, practicing mixed cropping and planting trees to provide for fodder and shade/ shelter for their crops. The negative effects of climate change on grain legume can be mediated by suitable strategies such as using stress tolerant varieties and adjusting time of sowing to favourable weather and mixed/inter cropping. Indigenous varieties which are drought resistant and stress tolerant can be grown.

Crop varieties resistant/tolerant to drought, flood, salinity, crop pests and other stress conditions are constantly being released. Such seeds are required to be supplied to identified areas. Production of such seeds in formal and informal system is necessary. At the same time conservation of plant genetic resources can be made by *in-situ* and *ex-situ* methods and village gene banks and seed banks. Plant breeders must develop crop varieties to adapt to climate change. Participatory plant breeding is another approach where farmers and plant breeders collaborate in development of crop varieties that are suitable for problematic areas.

(iv) Inclusion of legumes into cropping system: Legumes improve N levels in cultivated soil through biological N fixation and scavenging residual soil N, making it available for subsequent crops. Legumes are prolific leaf shedders, and act as a natural manure to conserve soil moisture and maintain soil fertility. Inclusion of legumes into the cropping system as a sole crop , as a mixed, relay or catch crop, as a cover and green manuring crop, in mixed cropping systems and as forage crops to improve livestock nutrition can be climate resilient.

Biological nitrogen fixation (BNF) is a complex mechanism which has been displayed in Fig. 9.12.

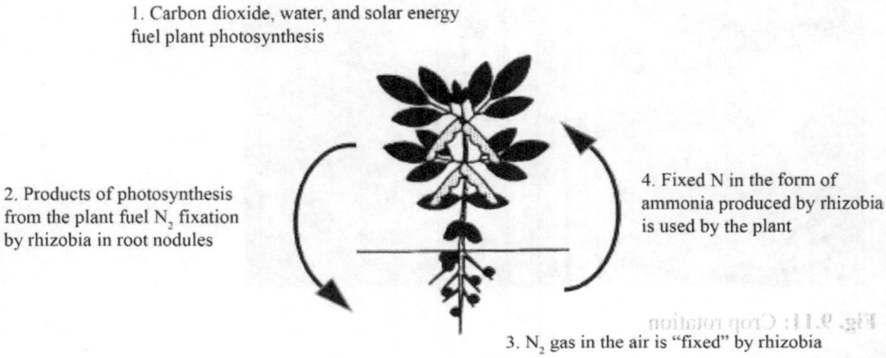

Fig. 9.12: BNF through legume-rhizobia symbiosis

Depending on many factors the amount of BNF varies from crop to crop as indicated in Table 9.3

Table 9.3: Estimated amount of nitrogen fixed by various legumes (FAO, 1984)

Plant	Scientific name	Nitrogen fixed Kg/ha/year
Horse gram	*Vicia faba*	45-552
Pigeon pea	*Cajanas cajan*	168-280
Cowpea	*Vigna unguiculata*	73-354
Mung bean	*Vigna mungo*	63-342
Chickpea	*Cicer arietinum*	103
Lentil	*Lens esculenta*	88-114
Peanut	*Arachis hypogaea*	72-124
Bean	*Phaseolus vulgaris*	40-70
Pea	*Pisum sativum*	55-77

(v) System of Rice Intensification (SRI): System of rice intensification (SRI) is the alternate wetting and drying of soil during the rice cropping season rather than continuous flooding. Flooded rice consumes more irrigation water and produces methane which is 25 times more potent than CO_2 as a GHG. SRI is a resource conserving system of cultivation. It gives better yield with less seed, less water and less nutrient under alternate wetting and drying system of rice cultivation. SRI reduces production of methane from rice field substantially and reduces consumption of water. However, it requires more manpower for cultivation which is compensated by low input cost. The benefits of SRI are:

- SRI increases crop productivity with low input use
- It produces productive and robust rice phenotypes from given plant genotypes.
- Under SRI crop is tolerant to biotic/abiotic stresses and it reduces GHGs from rice fields.
- The System enables farmers to adapt to and mitigate climate change.

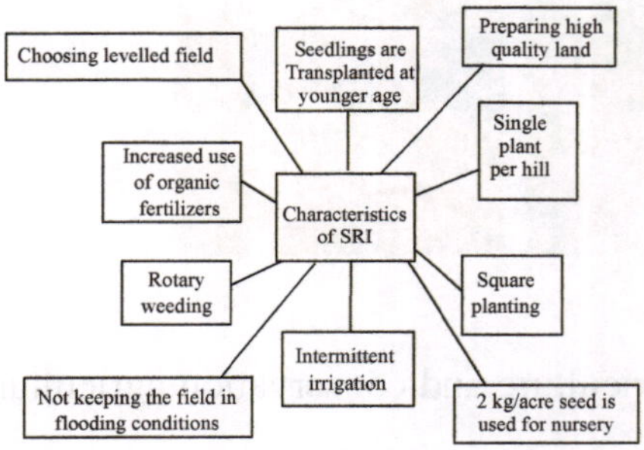

Fig. 9.13: Characteristics of SRI

Fig. 9.4 Tillering and rooting in SRI

Conservation agriculture: Conservation agriculture is an approach that combines limiting soil disturbance to a minimum, maintaining soil cover and diversifying crop production. It increases the resilience of the farming systems in a number of ways. The soil is kept fertile and protected from erosion and evaporation, water and nutrients are used efficiently, ground cover slows down the flow of water on the surface, allowing it to soak and infiltrate into the soil for crop use. Conservation agriculture reduces carbon dioxide emissions from tractor use, as the farm power requirements are lower and fewer passes across the field are needed. These 'carbon savings' exceed the carbon costs related to the use of chemical herbicides for weed control. Soil cover reduces moisture loss by evaporation, maintains soil temperature, reduces soil erosion, restores soil carbon by decomposition of crop residues and provides food for soil microorganisms. Crop rotation reduces pest and disease attack and replenishes soil nutrients. Minimum tillage increases population of beneficial organisms in soil which takes care of soil tillage to build up soil properties.

CA comprises three basic components:

- Reduced tillage
- Retention of crop residues on soil surface.
- Crop diversification (rotation, intercropping).

Fig. 9.15: Conservation agriculture

A comparison between conventional agriculture and conservation agriculture is given below.

Properties	Conventional agriculture	Conservation agriculture
Soil structure	Repeated tillage reduces soil organic matter, destroys soil structure and develops a hard pan.	Minimum tillage does not disturb soil structure and organic matter builds up soil.
Soil moisture	In absence of soil cover, evaporation is more and waterlogging develops due to hard pan.	Due to reduced evaporation, higher rate of infiltration and storage in soil, available moisture for crop is more.
Soil erosion	Crust formation and sealing of surface pores prevents water seepage and encourages erosion. In sloppy soil erosion becomes more.	Cover cropping and mulching protect the rainfall impact and reduces erosion. Roots bind the soil together which resists soil erosion.
Soil fertility	Ploughing breaks organic matter and kills micro- organisms which affects soil health. Monoculture removes valuable nutrients from soil.	Crop residues and cover crops add organic matter and addition of compost improves the soil fertility.
Pests and weeds	Repeated cultivation of single crop encourages weeds and pests to grow which require more pesticides.	Cover crops and mulch smoothers weeds and crop rotation reduce pest attack. A healthy soil helps control of pests and diseases.
Cost of labour	Conventional farming and mechanisation require more labour and fuel. Herbicides may reduce labour, but causes harm to the soil environment.	It requires more labour in the beginning year and labour requirement is reduces in following years.

Integrated nutrient management: Integrated nutrient management (INM) balances the use of organic matter obtained from the farm or from other sources with the judicious use of nutrients from mineral sources. It is a holistic approach that aims at higher yields and improvement in soil fertility.

Apart from applying organic manures and oil cakes, nutrient recycling by legume could be a potential component of integrated nutrient management. Green manuring with legumes like *Crotolaria* or *Sesbania* is a common practice in the rice growing areas of India. A 6-8 week old green manure crop of *Crotolaria* or *Sesbania* accumulates about 3-4 tonne dry matter/ha and fixes 100-120 kg N/ha. This can supplement up to 50 % of the total N requirement of rice after incorporation into soil.

There are two ways to manage nutrients: micro-dosing and fertilizer application strategies. Micro-dosing involves application of affordable quantities of fertliser to the planting holes at the time of sowing and top dressing after 3-4 weeks of emergence of seedlings. However, micro-dosing is time-consuming and laborious. Fertilizer application strategies include timing and placement of fertilizer applications which can influence fertilizer use efficiency and ultimately crop production. When a soil test indicates a particular nutrient deficiency, considerations of nutrient placement involve the type of fertilizer being applied, tillage and crop rotation practices, choice of crop, access to necessary equipment, nutrient mobility in the soil, and soil characteristics. Good management practices for nitrogen fertilization includes application of fertiliser on the basis of soil tests, application of fertiliser as per expected yield, and management of irrigation to avoid leaching loss. Phosphate fertiliser can be applied in bands close to root system. At low fertilization rates, the efficiency of P uptake by the crop is greater for banding than broadcast application, especially for P applications on soils with a high P-fixation capacity, such as organic soils. Banding of P fertilizer often decreases soil pH within the narrow application zone, which encourages improved P availability over a longer period of crop growth. This phenomenon is particularly important for soils with naturally high P-fixation capacities. Site specific nitrogen management is possible by using leaf colour charts and greenSeekers.

Carbon sequestration: Leaving the crop residues in soil without burning, minimum tillage and addition of compost can sequester carbon in the soil. For encouraging carbon sequestration, cover and catch cropping, restoration of degraded land, soil conservation and erosion control methods are essential. The establishment of deep-rooted and perennial plants will facilitate carbon sequestration, and trees can be used for live fencing, windbreaks, fodder as riparian forest buffers, etc.

Integrated Pest Management: Climate change often increases pest attack in crops. Indiscriminate use of chemicals may provide pest control for a short period, but can cause environmental problems like pest resurgence, pest resistance and endemic pest attack in the long run. Integrated pest management (IPM) is a broad ecological approach which involves the use of appropriate measures to discourage the development of pest populations and keep them within economic injury level. It is a system that relies on varietal, cultural, mechanical, physical, biological and need based chemical control tactics to maintain a natural balance in the crop ecosystem

The principles of IPM approach include growing healthy crops, understanding ecological processes in the fields and encouraging natural pest management mechanisms that maintain ecological balances among populations of pests and their natural enemies (predators, parasites, antagonists), observing fields regularly, and building farmers' capacity and understanding of ecological needs so that they are empowered to take the best pest management decisions in their own fields. Managing options for insect pests and diseases include environment-friendly preparation of cow urine-based botanical pesticide, crop rotation, mixed cropping and measures that are based on local resources, knowledge and skill. Indigenous products like cow urine, botanical pesticides and bio-pesticides are used to encourage the activities of bio-agents. Laboratory reared predators and parasites are released for control of crop pests. Safe chemicals are used against the target pests when the population exceeds the economic threshold level (ETL).

Integrated farming system (IFS)

Integrated farming system is a diverse system that includes crops, livestock or forestry which is integrated at various scales to enhance the efficiency and environmental sustainability of the different production methods. Integrated farming system uses different enterprises (Fig.9.16) where the by-product of one component is used as input in the other component. IFS provides additional adaptation options to climate change and produces diversified products like grains, vegetables, fruits, meat, egg and milk.

Fig. 9.16: Integrated farming system

Climate smart approaches

At the field level, there are a wide range of agricultural practices and approaches that can contribute to increased production while still focusing on environmental sustainability. These practices and approaches for crop production can provide adaptation measures and/ or mitigation benefits. With appropriate technical, institutional, socio-economic and policy infrastructure in place, there is a huge potential for crop management practices and approaches for mitigation of climate change.

(i) Preservation of genetic resources

Crops, animals and other genetic resources will need to adapt to the impacts of climate change. Breeding activities should ensure that genetic resources for food and agriculture are able to adapt to new production conditions. The characters important for climate change adaptation include: (i) capacity to tolerate high temperatures and droughts; (ii) fire resistance and tolerance, especially for trees; (iii) resistance or tolerance to diseases and insect pests; (iv) capacity to utilize scarce and poor quality feed and soil; (v) tolerance to lower water quality, especially for aquatic organism; and (vi) capacity to range over harsh terrain in search of feed and water in case of livestock etc.

(ii) Energy management

Fossil fuels are likely to meet the bulk of the primary energy requirements. The prices of such non-renewable energy is increasing over the years while the costs of renewable energy have been declining recently. With this trend the gap between energy needs and access to energy will be large. Agriculture and energy have always been closely interlinked over time. The energy generated by the agri-food system can be partially used in the food supply chain or exported outside the system.

In agricultural sector, energy use has increased in farm mechanization, boosted fertilizer production and improved food processing and transportation. Since prices for nitrogen fertilizers and other fossil fuel-dependent inputs are closely related to the price of crude oil, the agri-food systems dependent upon fossil fuels pose serious challenges to agricultural development, and food security in the future. Food losses occur at all stages of the supply chain. Saving food from losses is one way of conserving energy. conserving energy. The food and energy losses has been displayed in Fig.7.4 in Chapter-7.

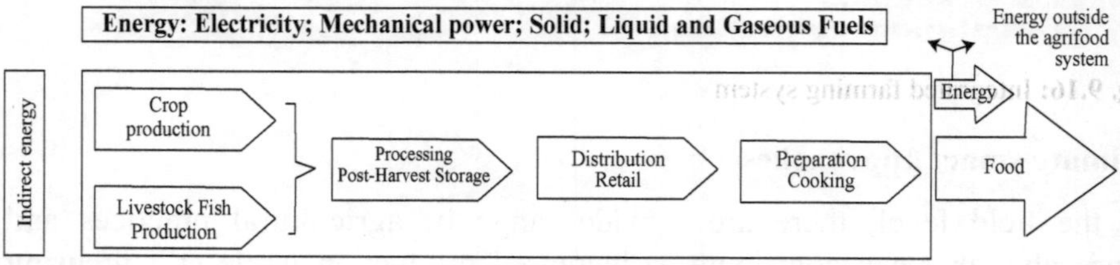

It is a challenge to develop a food production system that can emit less GHG and be resilient to fluctuating energy prices. This calls for an energy smart food system which must (i) improve energy efficiency at all stages of the agri-food chain; (ii) use diverse energy sources with an emphasis on renewable energy and contribute to renewable energy production through integrated food and renewable energy production; and (iii) require improved access to modern energy services. Bio-energy is used primarily for cooking and heating which may jeopardize food security and harm the environment, if not well managed.

The energy sector, which produces nearly 60 per cent of carbon dioxide (CO_2) emissions, is the largest contributor to climate change (FAO, 2011a). Globally, primary farm and fishery production accounts for around 20 per cent of the total energy demand for food, but produces 67 per cent of the GHGs (FAO, 2011 a). With high and volatile fossil fuel prices, energy-smart food systems which improve access to modern energy services and increase energy diversity can contribute to energy security. Dependence on local energy sources can increase incomes and expand the diversity of energy sources. Use of biogas can help ensure self-reliance in household energy which can reduce the amount spent on fuel-wood and chemical fertilizers. The energy supply needs to be 'climate-proofed' as much as possible to ensure that energy use in the agri-food system can be climate-smart. Precision agriculture, including a more efficient use of fertilizer, will lower CO_2 and N_2O emissions and reduce the consumption of fossil fuels.

Inefficient use of nitrogenous fertiliser leads to losses that are usually above 50 per cent and sometimes can amount to 60-70 per cent of applied nutrients

(Cassman et al., 2002). Increasing the efficiency of fertilizer application can result in optimal plant growth with minimal inputs would significantly improve the energy balance of food production.

The single most energy-consuming operation in a cropping cycle is soil tillage for land preparation, particularly ploughing. Minimum tillage or no-till system has become particularly attractive in times of high energy costs. Practices, such as zero tillage used in conservation agriculture, have the potential to bring about significant energy savings that can even reach up to 40-50 per cent (Doets et al., 2000; SCCA, 2012). For grain drying, modern continuous flow dryers can be used with much lower levels of energy than conventional dryers. Reduced energy use can also be achieved through the insulation of dryers, the re-circulation of heat recovery of out-going air and improved instrumentation and automatic control. Solar power (photovoltaic or solar heaters), wind and geothermal energy are all sources of energy particularly suitable for remote rural areas (Fig.9.17).

Fig. 9.17: Solar irrigation system

9.2.4 Climate smart Livestock & Fishery

Climate change affects the livestock through increased temperature, changes in the amount of rainfall and shifts in precipitation patterns. Indirect impacts are modifications in ecosystems, changes in the yields, quality and type of feed crops, possible increases in animal diseases and increased competition for resources. At the same time, livestock food chains are major contributors to greenhouse gas (GHG) emissions (FAO, 2006a).

The main strategies for climate smart livestock system are: (i) improved grazing system through rotational grazing to reduce CH_4 emissions and CO_2 removals through soil carbon sequestration, (ii) manure management, (iii) pasture management and nutrition, (iv) animal breeding to select more productive animals with lowering CH_4 emissions and more resistance to climate change,

(v) better nutrition and health management,(vi) early warning systems, (vi) agro-forestry practices, and (vii) mixed farming systems.

Fig. 9.18: Grazing management

Ecosystem approach to fisheries and aquaculture would integrate fisheries, aquaculture, agriculture and forestry. In rice-fish culture in flooded rice fields there is synergetic increase in production of rice and fish. The fish are harvested by reducing the water level, forcing them to swim to the lowest-lying part of the fields, where they can be caught easily. The agricultural producers get two harvests – rice and fish – from the same land, improving income and food security.

9.2.5 Climate smart forestry

Forests and trees provide a range of forest products and ecosystem services. Mainstreaming climate change into forest policy and practice will entail capturing synergies and managing trade-offs with other forest management objectives. Climate smart agriculture in the forest sector will entail designing adaptation actions targeted at the most vulnerable communities and sectors of the population and forest systems. Climate change responses, particularly mitigation policies, are leading to changes in markets for forest products and may drive substitution of forest products for more energy- or carbon-intensive products. Forest planning and management will have to be sensitive and adjust to such market changes.

World Agroforestry (ICRAF) defines agroforestry as "a farming system that integrates crops and livestock with trees and shrubs." Agroforestry practices such as use of shade trees are another example of an integrated practice. The resulting biological interactions in the agro-forestry system provide multiple benefits, including diversified income sources, increased biological production, better water quality, and improved habitat for both humans and wildlife. Agroforestry has multiple benefits such as increase in carbon sequestration, higher income from non-wood and timber, controlling soil erosion and maintaining soil fertility. Agroforestry brings a number of benefits in terms of soil carbon.

9.2.6 Biodiversity conservation

Bio-diversity is especially important in a climate-smart approach because it contributes to pest and disease management, which has direct effect on yields and revenues and can be very costly and labour intensive. Agriculture and allied agriculture production systems depend on three components of bio-diversity conservation such as (i) the diversity of the species, (ii) the diversity within each species, and (iii) the diversity of eco-systems. Greater diversity of crops and other living organisms in a cropping system is required to ensure farm resilience, economic stability and profitability. Integrating multipurpose crop varieties, which can improve the functional and productive management on the farm and are climate-smart.

References

Beddington, J., Asaduzzaman, M., Clark, M., Fernandez, A., Guillou, M., Jahn, M., Erda, L., Mamo, T., Van Bo, N., Nobre, C.A., Scholes ,R., Sharma, R. & Wakhungu, J. 2012. Achieving food security in the face of climate change. Final report from the Commission on Sustainable Agriculture and Climate Change. Copenhagen.

Bergström, L. 1995. Leaching of dichlorprop and nitrate in structured soil. Environmental Pollution, 87: 189–195.

Brown, S. & Lugo, E. 1990. Tropical secondary forests. Journal of Tropical Ecology, 6: 1–32.

Cassman, K. G., Dobermann, A. & Waters, D.T. 2002. Agroecosystems, nitrogen use efficiency, and nitrogen management. Ambio, 31: 131—140.

Davies, D.W., Garwood, T.W.O. & Rochford, A.D.H. 1996. Factors affecting nitrate leaching from a calcareous loam in East Anglia. Journal of Agric. Science, 126: 75–86. Cambridge University Press.

Doets, C.E.M., Best, G. & Friedrich, T. 2000. Energy and conservation agriculture. Internal FAO paper, Rome. Energy Sector Management Assistance Program (ESMAP). 2011. Climate impacts on energy systems: key issues for energy sector adaptation. World Bank. (available at https://www.esmap.org/sites/esmap.org/files/ Document Library/E-book_Climate %20Impacts %20on%20 nergy%20Systems_BOOK_resized.pdf)

FAO & National Drought Mitigation Centre (NDMC). 2008. The Near East drought planning manual: guidelines for drought mitigation and preparedness planning. Food and Agriculture Organization of the United Nations Regional Office for the Near East, Cairo and University of Nebraska-Lincoln, National Drought Mitigation Centre

FAO (2013) Climate-smart agriculture sourcebook. Rome. 558 pp. http://www.fao.org/ docrep/018/i3325e/ i3325e.pdf

FAO. 2006a. Livestock's long shadow: environmental issues and options. Rome. (available at ftp://ftp.fao.org/ docrep/FAO/010/A0701E/A0701E00.pdf)

FAO. 2007. Modernising irrigation management – the MASSCOTE approach. FAO irrigation and drainage paper No. 63. Rome.

FAO. 2011. Save and grow: a policymaker's guide to the sustainable intensification of smallholder crop production. Rome.

FAO. 2011a. "Energy-smart " food for people and climate. Issue paper. (available at http://www.fao.org/docrep/014/i2454e/i2454e00.pdf)

FAO. 2011b. The case for "energy-smart" food systems. Policy Brief. (available at http://www.fao.org/docrep/014/i2456e/i2456e00.pdf)

FAO. 2012. Coping with water scarcity: an action framework for agriculture and food security. Water report No. 38. Rome.

FAO. 2012a. Economics of plant genetic resource management for adaptation to climate change. A review of selected literature, by S. Asfaw & L. Lipper. (available at http://www.fao.org/docrep/015/an649e/an649e00.pdf)

FAO. 2012b. FAO statistical yearbook, 2012: world food and agriculture. (available at http://www.fao.org/docrep/015/i2490e/i2490e00.htm)

FAO-PAR (Platform for Agro-biodiversity). 2011. Biodiversity for food and agriculture: contributing to food security and sustainability in a changing world. Rome. (available at http://agrobiodiversityplatform. org/files/2011/04/ PAR-FAO-book_lr.pdf)

Farage, P., Ardö, J., Olsson, L., Rienzi, E., Ball, A. & Pretty, J. 2007. The potential for soil carbon sequestration in three tropical dryland farming systems of Africa and Latin America: a modelling approach. Soil and Tillage Research, 94(2): 457–472.

Gors, M.J., Howsc, P.W., Lane, P.W., Christian, D.G. & Harris, G.L. 1993. Losses of nitrate-nitrogen in water draining from under autumn-sown crops established by direct drilling or mouldboard ploughing. Journal of Soil Science, 44(1): 35–48. (available at http://onlinelibrary.wiley.com/doi/10.1111/j.1365-2389.1993.tb00432.x/

Jiang, Q.Q. & Bakken, L.R. 1999. Comparison of Nitrosospira strains isolated from terrestrial environments.FEMS Microbiol Ecol., 30: 171–186.

Lindau CW, Bollick PK, DeLaune RD, Moiser AR, Bronson KF (1993) Mithane mitigation in flooded Lousiana rice fields. Bio Fert Soila : 15 (3): 174-178

Macdonald, A.J., Powlson, D.S., Poulton, P.R. & Jenkinson, D.S. 1989. Unused fertiliser nitrogen in arable soils-its contribution to nitrate leaching. Journal of the Science of Food and Agriculture, 46: 407–419.

Noellemeyer, E., Frank, F., Alvarez, C., Morazzo, G. & Quiroga, A. 2008. Carbon contents and aggregation related to soil physical and biological properties under a land-use sequence in the semiarid region of Central Argentina. Soil & Tillage Research, 99: 179–190.

Soil Conservation Council of Canada (SCCA). 2012. Producers are finding the reduced tillage zone. (available at http://www.soilcc.ca/ggmp_region/nb-02.php)

Sparkes, J. & Stoutjesdijk, P. 2011. Biochar: implications for agricultural productivity. ABARES technical report, 11 (6). Australian Bureau.

Turral, H., Burke, J. & Faurès, J.M. 2011. Climate change, water and food security. FAO Water Report No. 36. Rome, FAO. (available at http://www.unwater. org/downloads/unw_ccpol_web.pdf

World Agroforestry Centre (ICRAF). 444 pp. https:// www.worldagroforestry.org/output/climate-smart-landscapes-multifunctionality-practice

Zach, A., Tiessen, H. & Noellemeyer, E. 2006. Carbon turnover and carbon-13 natural abundance under land use change in semiarid Savanna soils of La Pampa, Argentina. Soil Science Society of America Journal, 70(5): 1541–1546.

10

Climate Smart Village

10.1 Application of Climate-Smart Agriculture

Agricultural production systems are expected to change in response to climate change and ensure food security in different countries. Climate variability may further affect the agricultural production in future. To meet such challenges, there are many options to reduce the negative impacts of climate change on agricultural systems, make them resilient to climate change, and reduce emissions. Many technologies and practices can increase crop yields, farm income, and input-use efficiency, and may reduce GHG emissions (Khatri Chhetri et al. 2017). Climate-smart agriculture (CSA) is suggested to address the adverse effects of climate change and ensure food security. It is intended to achieve the three objectives of ensuring productivity, adapting and building resilience and reducing emission of GHG.

10.2 Concept of Climate Smart Village (CSV) Approach

The CGIAR Research Programme on Climate Change, Agriculture and Food Security (CCAFS) has developed the Climate-Smart Village (CSV) approach as a means to agricultural research for development (AR4D) in the context of climate change. The key features of AR4D approach are:

- Sites for testing, through participatory methods, technological and institutional options.
- Sites where climate change in its broadest context is considered, but in relation to local realities: long-term adaptation, avoiding maladaptation, climate risk management, low emissions development.
- Embodies a holistic vision for climate change action – not a silver bullet approach.
- A platform for socially inclusive, multistakeholder collaborative work.
- Founded on the principle of bringing CSA to scale.
- Links global and local knowledge.

CSA is seen as practices, technologies, services and institutional options in a broad sense. The CSV AR4D approach considers the following components in proposed sites (Fig.10.1). In AR4D approach, CSVs are multi-stakeholder learning platforms, participatory test-beds for generating evidences for further scaling.

The decision support to farmers and other stakeholders provided by an ideal CSV approach are : (i) village/community agricultural land use plans and contingency plans considering current and future climate risks, agro-ecological and socioeconomic conditions, and markets; (ii) portfolios of agricultural practices, technologies, and programmes that do not become maladaptive in future climate and market scenarios are assessed using models; (iii) strategic guidance before and during the planting season on the most suitable CSA practices, technologies, services, processes, and institutional options considering market and resource availability such as capital, labour, and markets done in a participatory mode with local farmer groups such as farmers' self-help groups(SHGs), water-users associations (WUA), local extensions systems, or local private-sector actors; (iv) tactical guidance to farmers on using real-time weather forecasts and value-added information and communication technology (ICT)-based agro-advisories; on accessing good quality inputs and technologies for improving water/ nutrient/energy use efficiencies, and on risk transfer through insurance mechanisms in case of crop and livestock losses; and (v) policy-level guidance on barriers and options to unlock CSA with local and national development, including consideration of financial needs to drive scaling. The CSV approach differs from existing agricultural development approaches, which usually depend on substantial external funding.

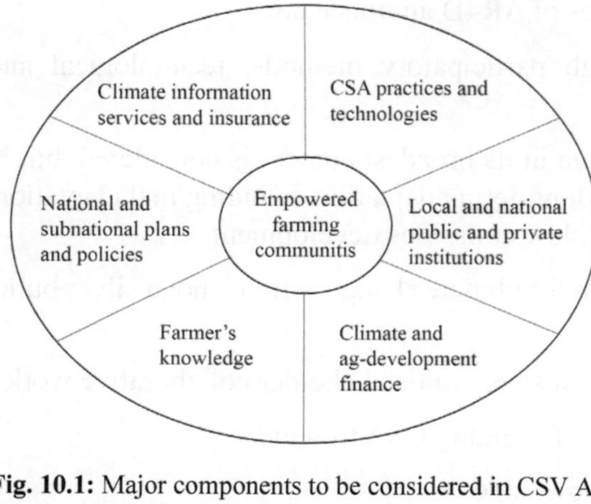

Fig. 10.1: Major components to be considered in CSV AR4D sites.

The CSV approach promotes adaptation and transformation options locally with capacity building. The approach aims to have positive impact on agriculture-dependent communities.CSV AR4D sites could be cluster of villages or landscapes with one or many villages.

10.2.1 Broad Objectives

Food security: Since climate change affects four dimensions of food security (availability, access, absorption and stability) the major objective of CSV is to ensure food security by adopting sustainable agricultural practices.

Adaptation: The CSV seeks to increase the capability of the stakeholders by building social capital, economic resources, awareness & training and adopting climate resilient technologies and developing infrastructure against climate change.

Mitigation: Mitigation strategies to reduce the emission of GHG from agriculture and allied sectors are attended in CSV.

10.2.2 CSV Interventions

The interventions include weather-smart, water-smart, seed/ breed smart, carbon/ nutrient-smart and knowledge smart (Fig.10.2).

Climate smart	Interventions
Weather smart	Weather forecasts, climate informed agro-advisories, weather insurance, climate analogues as a tool for forward planning, strategies to avoid maladaptation.
Water smart	Aquifer recharge, rainwater harvesting, community management of water, laser-land levelling, micro-irrigation, raised-bed planting, solar pumps.
Seed and breed smart	Adapted varieties and breeds, seed banks, including community-based activities.
Carbon/nutrient smart	Agroforestry, minimum tillage, land use systems, use of leaf colour chart, greenSeeker, livestock management, integrated nutrient management, biofuels.
Energy smart	Bio-fuel, fuel efficient engine, residue management, minimum tillage, solar solutions.
Knowledge smart	Cross sectoral links; local institutions, including learning platforms or farmer-to-farmer learning and capacity development, contingency planning, financial services, market information, gender equitable approaches, and off-farm risk management strategies.

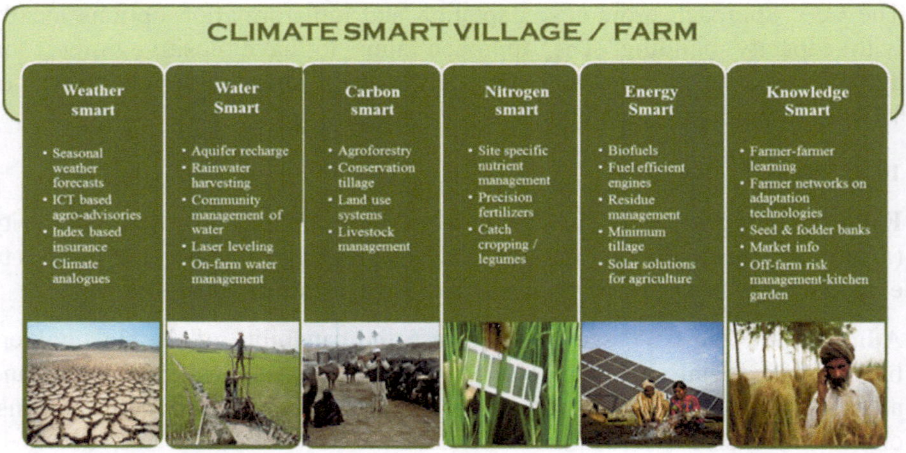

Fig. 10.2: Climate-smart interventions

Fig. 10.3: Use of leaf colour chart

Fig. 10.4: Use of green Seeker

10.2.3 Implementation of CSV

Effective implementation of climate smart agriculture requires an integrated approach in which science, technology, and decision making interact with local socioeconomic conditions and cultures (Steenwerth et al. 2014). A CSV is a platform for action research to test, develop, and support the scaling of CSA technologies and approaches to help in deriving practical adaptation options to ensure food security, improve resilience, and decrease emissions of greenhouse gases. The CSVs are meant to generate and scale CSA technologies and social learning approaches or to scale the CSV methodologies. It aims at identifying the best climate smart agricultural options to draw out lessons for policy makers and agricultural practitioners. The testing is done through a multi-stakeholder collaborative platform at CSV sites like cluster of villages, small landscapes, or 10 km^2 grids as every CSV site has its own theory of change linked to national priorities to ensure that it is consistent with initiatives and actions across different scales.

Fig. 10.5: Climate smart village

The strategies of the CSV approach are to:

- Understand the effectiveness of a variety of CSA options (practices, technologies, services, programs, and policies) not only to enhance productivity and raise incomes, but also to build climate resilience, increase adaptive capacity, and wherever possible, reduce GHG emissions;
- Develop solutions in anticipation of future climate change impacts;
- Understand the socio-economic, gender, and biophysical constraints and enablers for adoption; and
- Test and identify successful adoption incentives, finance opportunities, institutional arrangements, and scaling out/up mechanisms while ensuring alignment with local and national knowledge, institutions, and development plans.

The strategy of the CSV approach is not only to enhance productivity and raise incomes, but also to build climate resilience and reduce emission of GHG.

There are many options and they differ based on the CSV site, its agro-ecological characteristics, level of development, and capacity and interest of the farmers and of the local government. An ideal CSV approach provides five types of decision support to farmers and other stakeholders:

(a) Village/community agricultural land use plans and contingency plans considering current and future climate risks, agro-ecological and socio-economic conditions, and markets.

(b) Portfolios of agricultural practices, technologies, and programs that do not become maladaptive in future climate and market scenarios are assessed using models.

(c) Strategic guidance before (where feasible based on seasonal forecasts) and during the planting season on the most suitable CSA practices, technologies, services, processes, and institutional options considering market and resource availability such as capital, labour, and markets. This is done in a participatory mode with local farmer groups such as farmers' self-help groups, water-users associations, local extensions systems, or local private-sector actors.

(d) Tactical guidance to farmers on using real-time weather forecasts and value-added information and communication technology (ICT)-based agro-advisories; on accessing good quality inputs and technologies for improving water/nutrient/energy use efficiencies.

(e) Policy-level guidance on barriers and options to unlock CSA with local and national development, including consideration of financial needs to drive scaling.

The key focus in a typical CSV is on strengthening capacities and empowering the farming communities and their local organizations. In the process, partners in the CSV approach commit only limited resource support to catalyse the process.

10.3 Key Steps of the Climate-smart Village Approach Implementation

The CSV approach consists of four steps, guides adaptation planning and on-the-ground implementation with the objective of transforming farming communities to be more climate- smart.

(i) Baseline assessment, including climate risk analysis and gender and social inclusion analysis: Historical climate data are analysed to assess risks such as the probability of sowing failure due to inadequate rainfall at the start of the season. It also includes risks of wet and dry spells, and periods of heat or cold stresses during the cropping season. Future climate scenarios are analysed to assess the likely change in risk profiles and long-term suitability of the main cropping and livestock systems. Secondary data are used to assess GHG emissions from agriculture and simple calculators in those regions where mitigation is relevant.

In baseline assessment, the natural resources (land, water, and soil) and physical resources (infrastructure such as irrigation, technology, and markets) availability are assessed. It also includes assessment of socio-economic resources such as human (labour and education), financial (source and amount of revenue and budget), and social (institutions and networks) resources. Local communities help guide the CSV design process, which includes their knowledge of climatic risk management.

A quick participatory assessment is made of the various practices and technologies being used in the village/region, as well as the potential ones that can be adopted, guided by agronomic studies in the region or homologous regions elsewhere. Availability of climate information services, such as weather-based agro-advisories and agriculture insurance schemes in the region are also assessed.

Assessment of current adaptation policies and institutions at different scales is essential to identify the potential for cooperation with and leveraging support from other entities. These assessments indicate the scope for mainstreaming the CSV approach into local adaptation policies and plans including other development initiatives. An evaluation is also done by gender and age to understand the different perspectives, division of labour, decision-making patterns, opportunities, and capacities in households and communities (Quisumbung and Pandolfelli 2010, Doss and Kieran 2013, Perez et al. 2015).

(ii) CSV design: Identification and prioritization of climate-smart technologies, practices and services based on biophysical, socio-economic, gender, policy and institutional context and possible synergies and trade-offs amongst individual activities are considered for CSV design. This step primarily focuses on developing a portfolio of practices and technologies dealing with food security, adaptation, and mitigation and on climate-information services that need to be initially piloted in the CSVs. The portfolio includes the six climate smarts already explained earlier in this chapter.

Strategic decision making in the CSV approach is a participatory activity. Proposed interventions must be relevant to the conditions defined by the baseline and show promise for wider scale relevancy beyond the CSV. The CSA practices, technologies, and services included in the portfolio for any site are the outcome of group discussions with women and men farmers, local government officials and researchers. Various tools—such as choice experiments, indicators-based ranking methods, and simple stakeholders' prioritization—are used for this exercise. The CSA portfolio may be influenced by gender difference in awareness levels, as well as by the choice of CSA technologies by the stakeholders. Several CCAFS projects have published details of how CSV portfolios of climate-smart interventions have been selected and designed (Taneja et al. 2014, Corner-Dolloff et al. 2015).

(iii) Creating evidence: Evaluation and development of portfolios of climate-smart interventions are done to create evidences. This is done in two steps. First, a narrow range of interventions/portfolios is evaluated in a multi-stakeholder research platform (including CGIAR, national agricultural research systems, NGOs, and private-sector players, farmers and their institutions) established in CSV sites. Assessment of benefits, synergies and technologies is made from the individual farmers and community. The detailed evaluation is made by surveys, farmers' group evaluations and ICT based feedback system.

Modelling is often used to supplement the results of on-farm evaluation, especially to estimate the benefits of resilience and mitigation benefits through time. In the second step, the final portfolios are evaluated by farmers on a larger scale. Crowd-sourcing of experiences is used in order to understand farmers' preferences and technology adaptation domains across CSV sites in the region. Due consideration is also given on the identification and evaluation of constraints and barriers to adoptions of interventions along with ways to overcome them.

(iv)Scaling: This involves scaling up through policies and institutions, and scaling out to large areas through farm-to-farm and ICT-based approaches. After generation of evidences the promising innovation are scaled up. Scaling mechanisms tested across the regions include:

- Horizontal scaling (scaling out) of climate-smart options: CSVs provide demonstration sites for farmer-to-farmer learning (often through self-help groups or producer organizations) and/or enable local promotion of CSA options through local government plans, programs, and policies or through private-sector business models.
- Vertical scaling (scaling up): CSV research and lessons learned provide evidence for the efficacy of practices, technologies, services, processes, and institutional options and are thus able to: influence large-scale

Horizontal scaling out is done from farmer to farmer through networking using ICT-based tools, farmers' fairs, farmer exchange visits, and participatory videos of the adopted technologies, or through engagement with local government institutions. Connecting with women's organizations is another effective scaling-out approach. Institutional and financial needs are determined, for both local and national (or subnational) contexts, often using appropriate business models. The potential for consolidation and convergence of current agricultural development schemes/plans in the region is explored. Consideration of national and regional gender machineries and social development programme also helps integrate social inclusion concerns into the CSVs as well as to orient social development programme to climate change impacts and action.

Monitoring and evaluation provide feedback to the CSV design process, which is based on learning and adjustment. Finally, adaptation pathways for the CSA approach—including technology and services portfolios, capacity building programmes, institutional and financial mechanisms, and potential costs and benefits of programme implementation—are presented to policy makers for scaling up through replication in other sites, or provision of incentives for adoption through large-scale agricultural investment plans. Although conceptually this step is placed at the end of the sequence, it can vary according to the CSV context.

The four steps are displayed in Fig. 10.6.

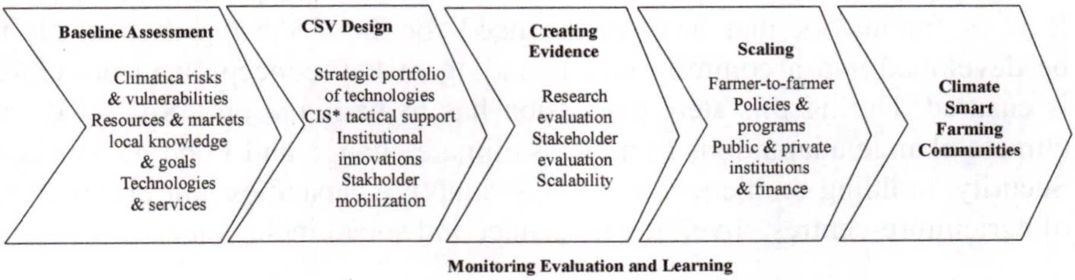

Fig. 10.6: Key steps in CSV approach

The CGIAR Research Programme on Climate Change, Agriculture and Food Security (CCAFS) and the International Institute of Rural Reconstruction (IIRR) have published a guide for Agricultural Development Workers and policy makers on CSV approach where eight steps have been suggested (Fig.10.7).

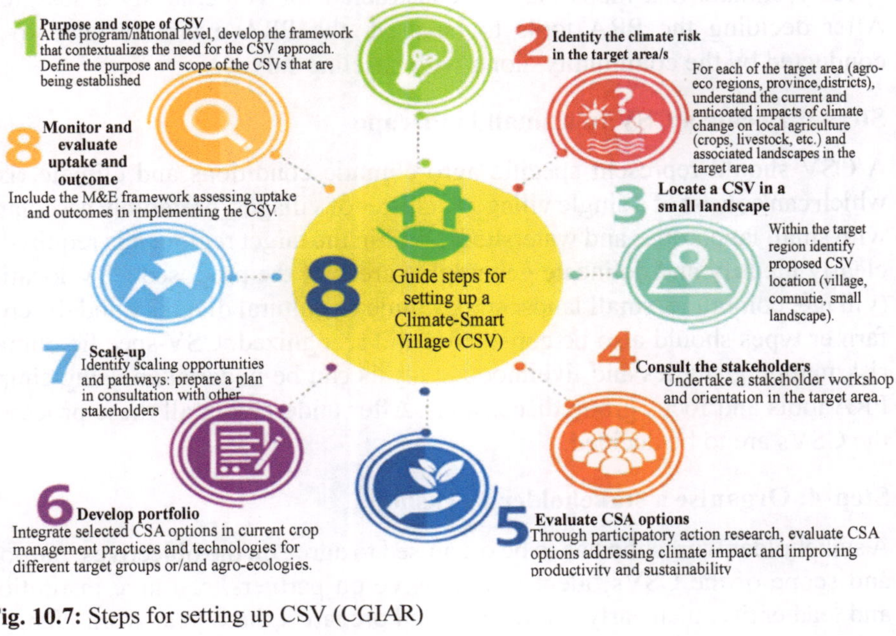

Fig. 10.7: Steps for setting up CSV (CGIAR)

Step-1: Define the purpose and scope of the CSV/s that is/are being established

In the first step discussion has to be made on climate change impacts, climate-smart agriculture concepts and practices including key components and elements of Climate-Smart Village approach and ways to achieve multiple outcomes - resilience building, increasing adaptive capacities and productivity, reducing gender inequality, and social inclusion. At the programme/national level, a framework that justifies the need for the CSV approach needs to be developed and a common understanding of key concepts and principles is ensured. During this step, discussion has to be made on CSA, CSV and climate change adaptations to address climate change and Food & Nutrition Security, building resilience, increasing adaptive capacities and productivity of agriculture-centres, livelihoods, gender and social inclusion.

Step-2: Identify the climate risk in the target area/s

In order to understand the climate risks and vulnerabilities using available assessment tools like Participatory Rural Appraisal (PRA) will be used in the target area (agro-ecological regions, districts). The current and anticipated impacts of climate change on local agriculture (crops, livestock, etc.) and the associated landscapes can be understood by conducting rapid baseline study, household-level assessments, and focus group discussions etc. Location specific climate risk maps may be constructed for vulnerability assessment. After deciding the PRA tools to be used, the PRA exercises have to be conducted by the community along with baseline survey.

Step-3: Locate a CSV in a small landscape

A CSV should represent specific agro-climatic conditions and climate risks which can consist of a single village or cluster of villages, preferably associated with small landscapes and watersheds. Within the target region, it is required to clearly identify and delineate geographic areas of the proposed CSV location (village, commune, small landscape). Ethnic or cultural diversity and different farmer types should also be considered and recognized. CSV-specific climate risk mapping studies and livelihood analysis can be undertaken using simple PRA tools and focus group discussions. After undertaking all these processes the CSVs are to be decided.

Step-4: Organise a stakeholder workshop

A stakeholder workshop has to be organised to agree on the objectives, purpose and scope of the CSVs; identify and agree on partners, resource institution, and lead entity; plan early for scaling; and prepare a three-year work plan.

Step-5: Evaluate CSA options

Through participatory action research, evaluate CSA options addressing climate impacts and improving productivity and sustainability. In evaluating CSA options, recognize that CSVs are platforms for testing, deriving, and subsequently promoting locally-adapted options in out scaling efforts. It is required to prioritise options unique to local context in CSV.

Step-6: Develop portfolio

CSA options in crop management practices and technologies for different target groups and agro-ecological situations need to be integrated. While developing a CSA portfolio the climate risks and livelihoods and expectations of target groups must be considered. During the actual field visit, ensure there is exchange of ideas between/among participants and CSV stakeholders. At the end of each site visit, facilitate reflection sessions. The key lessons are to be synthesised and farm-level adaptation strategies are to be identified using field-derived lessons and emerging ideas.

Step-7: Identify scaling opportunities and pathways

In consultation with the stakeholders a plan has to be prepared to scale-up, identify opportunities for horizontal and vertical scaling, organize events to share CSA technologies and tools and to disseminate findings, and engage media and communications specialists in promoting successes and stories.

Step-8: Monitor and evaluate (M&E) uptake and outcome

The required activities are to implement M&E of uptake and outcome; conducting (on annual basis) local reviews and assessment; assessing outcomes from action research undertaken in CSV on a regular basis using multidisciplinary teams, multiple stakeholder representatives, and community representatives; using indicators such as productivity, livelihoods, food and nutrition equity, gender, and resilience to monitor progress; and developing outcome stories (substantiated by evidence and demonstrated by the scale of adoption, with the likelihood of being sustained).

Participatory Vulnerability Assessment (PVA)

PVA is an approach that combines various tools and methods that communities can use to generate and analyse information enabling them to identify options to address current climate change impacts and adaptation options. In practice, PVA outputs include but not limited to community profile (village history, current state of natural resources, land use, tenurial arrangements), livelihood profile, climate change perceptions and coping mechanisms (perceived climate change impacts). Data gathering methods may include focus group discussion with PRA tools, key informant and group interviews and review of secondary data and climate-related documents.

Participatory Action Research (PAR)

PAR, an approach that brings together farmers, researchers and resource institutions, can produce sustainable solutions to address climate change impact in agriculture. PAR as defined by CCAFS is a "collaborative approach to research, education and action, and helps gather information in order to solve social or environmental issues". PAR in the context of resilient building in agriculture is used to test, identify and evaluate technological options and processes. Unlike conventional action research, testing and development of technologies are done at the farmers' field. These technological options, lessons and new knowledge can be scaled out through indigenous and emerging pathways.

Situational Analysis and Needs Assessments

In order to make a situation analysis the following information are to be gathered by using participatory methods such as focus group discussion (FGD), Key Informant's Interview (KII), and collection of information from secondary sources.

- Geographical location
- Demographics
- Local climatic information
- Natural resource utilization (land use, water resources, mining)
- Production and livelihood systems
- Food security status and trends
- Hazards and vulnerability
- Mitigation measures
- Climate change perception
- Climate change impacts
- Institutional landscape and governance
- Natural resource management initiatives
- Organizational landscape

- Information network
- Social and gender differentiation
- Health/Nutrition profiles and other livelihood outcomes.

Participatory Rural Appraisal Tools (PRA tools)

Participatory Rural Appraisal that facilitates analysis of community situation and problems is undertaken by the community members themselves. Selected PRA tools to be used are outlined below.

Name of PRA tool to be used	Purpose
Community mapping	To identify the community's boundaries, road networks, river and springs, landmarks and infrastructure and houses.
Historical time line (time trend)	To take stock of the significant events like major disasters, socio-economic and political events, changes on land scape and natural resources. It also shows the trends, frequency and intensity of events.
Seasonality calendar	To depict pattern of climate, livelihood activities, occurrence of pests and diseases, health issues, social and cultural activities in different months of the year.
Livelihood matrix	To list the various work performed and identify whether each activity is dominantly done by men or women. It also shows who will be most likely affected if a climate hazard may impact the livelihood.
Problem tree	To identifies the three major problems and challenges on livelihood and their different factors and reasons. Information and data gathered are rooted mainly on developmental issues and problems which are worsen by climate hazards and changes in climate patterns.
Vulnerability mapping	To identify and list out vulnerable families, individuals and context analysis of socio-economic patterns which will be useful in understanding the patterns of vulnerability to multiple risks, including climate risks.
Livelihood analysis	To know how individuals, households and groups of households make their living and the access of men and women to resources and services.
Stakeholder analysis	To identify different stakeholders for every adaptation and mitigation activity proposed.

References

Aggarwal, P. K., R. Zougmoré, and J. Kinyangi. 2013. Climatesmart villages: a community approach to sustainable agricultural development. CGIAR Research Program on Climate Change, Agriculture and Food Security (CCAFS), Copenhagen, Denmark.

Corner-Dolloff, C., A. M. Loboguerrero, M. Lizarazo, A. Nowak, F. Howland, N. Andrieu, and A. Jarvis. 2015. Decision-support framework for targeting investment towards climate-smart agriculture practices and programs. *In* C. Field, J. Jouzel, and H. Le Treut, compilers. *Our common future under climate change, international scientific conference*, 7–10 July 2015, Paris, France.

Doss, C., and C. Kieran(2013). *Standards for collecting sex-disaggregated data for gender analysis: a guide for CGIAR researchers.* CGIAR Consortium, Montpelier, France. [online] URL: http://www.pim.cgiar.org/files/2012/05/Standards-for-Collecting-Sex-Disaggregated-Data-for-Gender-Analysis.pdf

Food and Agriculture Organization (FAO), 2015. The state of food insecurity in the world 2015. Strengthening the enabling environment for food security and nutrition. FAO, Rome, Italy. [online] URL: http://www.fao.org/3/a-i4030e.pdf

Khatri-Chhetri, A., P. K. Aggarwal, P. K. Joshi, and S. Vyas. 2017.Farmers' prioritization of climate-smart agriculture CSA technologies. Agricultural Systems 151:184–191. http://dx.doi. org/10.1016/j.agsy.2016.10.005

Perez, C., E. M. Jones, P. Kristjanson, L. Cramer. P. K. Thornton,W. Förch, and C. Barahona. 2015. How resilient are farming households and communities to a changing climate in Africa? A gender-based perspective. Global Environmental Change 34:95–107. http://dx.doi.org/10.1016/j.gloenvcha.2015.06.003

Quisumbing, A.R., and L. Pandolfelli. 2010. Promising approaches to address the needs of poor female farmers: Resources, constraints, and interventions. *World Development* 384:581–592. http://dx.doi.org/10.1016/j.worlddev.2009.10.006

Steenwerth, K. L *et al* .2014. Climate-smart agriculture global research agenda: scientific basis for action. Agriculture and Food Security 3(1): 11. http://dx.doi.org/10.1186/2048-7010-3-11

Taneja, G., B. D. Pal, P. K. Joshi, P. K.Aggarwal, and N. K. Tyagi. 2014. *Farmers preferences for climate-smart agriculture: an assessment in the Indo-Gangetic Plain.* IFPRI Discussion Paper 1337. New Delhi, India. URL: https://ccafs.cgiar.org/publications/farmers-preferences-climate-agriculture-assessment-indo-gangetic-plain#.WSZhT_mGOUk

11

Extension Approach to Climate Smart Agriculture

11.1 Services for CSA

Under the CSA projects and programmes, the services like supply of inputs and technology transfer through extension services are provided to the farmers. The inputs like seeds of crops are provided by the input dealers at the subsidized rate and seedlings of fruits and vegetables are provided by both private and government agencies to grow under general cultivation and poly-house cultivation technology. Similarly other services like soil testing facilities, issue of soil health cards, construction of micro-irrigation projects and rain harvesting structures are made available to farmers. Under insurance scheme, loanee farmers are supported to pay premium dues for availing crop insurance facilities. Other services like trainings at various levels, demonstrations, exposure visits, exhibitions, *kisan melas*, field days, monitoring and evaluation of all these activities under these programmes are provided to the farmers with collaboration of extension officers. The summary of services and service providers for the farmers are indicated in Table 11.1.

Table 11.1: Services and service providers to farmers

Component	Services	Service providers	Beneficiaries
Input	Supply of seeds, agro-chemicals, seedlings/ grafts, traps, water harvesting structures, farm implements, soil testing, supply of health cards, crop insurance etc.	Traders, input dealers, State Agriculture University, Agriculture Department, Soil Testing Laboratories, Soil Conservation wings, banks etc.	Farmers
Extension service	Training, demonstration, exposure visits, organising exhibitions, *kishan mela*, field days, farmer field schools, awareness camps, field monitoring and evaluation of programmes.	Agriculture Department, KVK, Line departments, State Agriculture Universities, State Agriculture Management Extension and training Institute (SAMETI) etc.	Farmers, SGHs/FIGs

11.2 Barriers in the Adoption of Climate-Smart Agriculture Practices

There are several barriers to adopt climate smart agricultural practices at field level. The important barriers which need to be removed are:

- Lack of space on the small land holdings for installing water harvesting structures particularly in rain-fed areas ;
- Lack of labour availability to carry out the construction of the irrigation structures and taking out various agronomic practices;
- Inaccessibility of the good quality inputs by the farmers from the input dealers and traders ;
- As agriculture is the secondary occupation for some of the farmers, so lack of interest acts as the barrier in taking out the CSA practices in the farmers' fields ;
- Another barrier is poor market facilities especially for fruits and vegetables grown under CSA programmes ;
- Lack of access to credit facilities for insurance of the important crops; and
- Another big barrier i.e., conflicts exist among the farmers regarding the decision for taking out the extension activities like demonstrations, training camps at village level which lead to social and political problems in that particular villages.

11.3 Role of Extension

The key objectives of agriculture extension to provide a system in which knowledge, skills and attitudes relating to farming practices are transmitted to the farming community through various channels and methods. The main aims are; (i) transfer technology and knowledge regarding improved practices in agriculture and allied sectors, (ii) ensure practical application of knowledge and technology in the farm, and (iii) improve living standards of farm families.

Choice of extension approach can influence the ability of extension services to contribute to food security and income, adaptation and resilience, and climate change mitigation. The most popular extension approaches used in CSA are: (i) climate awareness programmes /campaigns, exhibitions, (ii) climate trainings, (iii) climate workshops, (iv) plant health rallies, (v) farmers field schools, (vi) exposure visits, (vii) demonstration of new technologies, (viii) participatory crop planning, (ix) dissemination of climate resilient technologies, and (x) establishment of agro-clinics.

11.4 Role of Rural Advisory Services

The expected effects of world temperature change, that embrace higher temperatures, additional frequent extreme weather events, water shortages, rising ocean levels, ocean natural process, land degradation, the disruption of ecosystems, and therefore the loss of diversity, may seriously compromise agriculture's ability to feed the foremost vulnerable populations and impede progress towards the reduction of hunger, deficiency disease, and poorness (FAO, 2016a). It is required to enhance the capability of farmers to manage risks of climate change and adopt effective measures for adaptation and mitigation. Promoting climate-smart agriculture involves strategies and agricultural practices supporting climate smart agriculture. Rural advisory service serves as a link between farmers and new sources of information and tools. The approaches and methods to be used in rural advisory service are demonstrations, trainings, individual farm visits, training of lead farmers and staff, training to input service providers, group discussion, exposure visits, farmer field schools, plant clinics, ICT based messaging and field days etc. Some of the important methods are described below.

Climate Awareness Mass Media Campaigns: Due to limitations of field extension staff, extension can be more efficiently performed using mass media. The most preferred channels are radio, television, telephone, SMS, APPs etc.

Climate Training: Education/training of intermediaries/extension personnel is important to update their knowledge related to climate change, their impacts, and consequences or on different adaptation and mitigation strategies. There are various topics in which training can be given. Farmer to farmer trainings can be organised through farmers field schools (FFS).

Plant Clinics: Plant clinic is an innovative approach which plays a vital role in timely diagnosis assuring food security and ushering prosperity by providing timely diagnosis of pests and diseases including nutrient deficiency and recommending solutions. In true sense, mobile clinic implies, a mobile clinic van with basic tools, public address system, relevant films with experts/ plant doctors rushing to affected or desired area, beyond clinic premises to diagnose the problem in the affected area and offer solution to suppress or manage the problem. Though the major role of plant clinic lies in diagnostics and advisory, the activities of plant clinic extend beyond plant clinic, with emphasis on extension, working more closely with farmers and organizations involved in promoting food production.

Plant Health Rally Approach: Plant health rally aims to deliver customised crop health messages to farmers in locations where they assemble. The extension staffs are trained on the best methods to quickly and effectively deliver the message and to answer farmers' questions. The recommendations to farmers are based on up-to-date information and evidence from research. The rallies also enable staff from crop disease and pest surveillance areas to work and share knowledge with extension staff for the benefit of farmers.

It is an extension method for quickly raising awareness about major agricultural risks or threats on important crops, to promote the use of improved agricultural practices. They are usually held in public spaces by the local extension officers and are open to everybody. A plant health rally may be spontaneous, attracting people with a banner and other announcements, or may target farmers who have been specifically mobilized for the event.

Contingency Crop Planning: In a situation of weather aberration, contingency of growing another crop in place of normally grown crop arises. Depending upon the nature of weather change crop is selected. Contingency cropping is highly location specific due to variation in amount and distribution of rainfall. Crops have to be selected with suitable crop duration to coincide with the length of the crop growing season. Generally short duration crops or catch crops may suit the situation. Farmers with economic strength and motivation for high profits with some amount of risk can go for crops of long duration. The long duration crops with flexibility or elasticity in yield are more suitable in contingent cropping.

It is participatory in nature because the local knowledge of farmers about the crops or various agricultural practices and scientific knowledge of scientist both play important role in crop planning. It is prepared with the help of agro-meteorological department. Use of mobile phones, videos, radios etc. is done to address the issue of climate change by creating awareness among the farmers about the availability of different adaptation and mitigation strategies.

Climate Farmers Field Schools (FFS): The farmer field school (FFS) approach is an alternative extension method to top-down extension. The Farmer Field School (FFS) is a participatory, non-formal extension approach based on experiential learning that puts farmers and their demands at the centre (FAO, 2002).

A group of farmers (25-30) assemble in the field school once in a week. In groups of five they observe and compare two plots over the course of an entire cropping season right from sowing to harvesting. One plot is a control plot with conventional methods while the other demonstrates the possible "best practices". They critically observe the plant development, taking samples of insects, weeds and diseased plants, and constructing simple cage experiments or comparing characteristics of different soils. The groups make presentation basing on their observation at the end of the week in a plenary session. In this programme farmers are able to investigate a wide range of topics relating to field problems and solutions. At the end of the season, a typical FFS group holds a field day to show the result.

Farmer-to-Farmer Extension (F2FE): Farmer to Farmer (F2F) extension approach in agriculture is a systematic communication between farmers. This approach aims to strengthen the information flow through the farmers. As a generic term we use it as 'farmer trainer' to communicate the message to other farmers.F2F approach helps in building the effective, farmer-centred extension systems. It can reduce the extension cost and workload of extension functionaries. The approach empowers farmers as change agents and helps extend adoption because farmers are more willing to find out from their colleagues than from extension staff (Franzel et al., 2015). The farmers and local institutions (e.g. producer organisations or village leaders) should play a key role in selecting farmer-trainers and monitoring and evaluating them. This helps make the programmes more accountable to the community or groups that they serve

ICT Supported Network: Information and Communication Technologies (ICTs) play a crucial role as a medium of data and communication in global climate change awareness, adaptation and mitigation strategies. However, the supply and adoption of ICTs vary between areas, developed and developing countries, urban and rural areas, and within rural areas themselves. The use of mobile phones, videos, radios, etc. is done to deal with the difficulty of global climate change by creating awareness among the farmers about the supply of various adaptation and mitigation strategies.

The impact of various extension tools is shown in Fig.11.1.

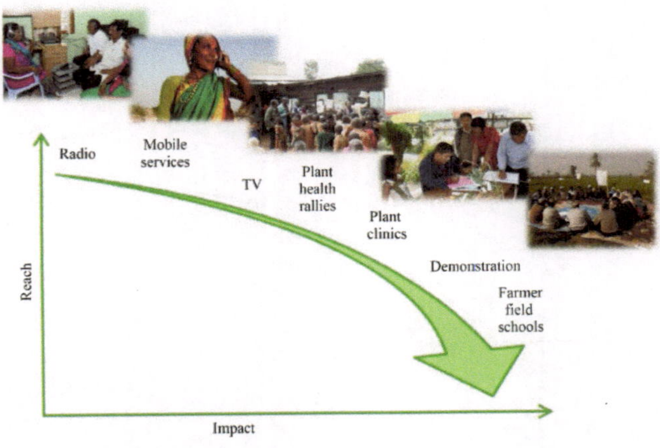

Fig. 11.1: Reach and impact of various extension advisory services

References

FAO 2002: From farmer field school to community IPM – Ten years of IPM training in Asia. Bangkok, Thailand. (FAO), Rome, Italy.

FAO. 2013a. Sourcebook on Climate-Smart Agriculture, Forestry and Fisheries, Food and Agriculture Organisation of the United Nations (FAO), Rome, Italy. Retrieved from http: //www.fao.org/climatechange/37491-0c425f2caa2f5e6 f3b9162d39c8507fa3.pdf

Franzel, S, et al. 2015: Farmer to-farmer extension. Good Practice Notes for Extension and Advisory Services. Lindau, Switzerland, Global Forum for Rural Advisory Services. Retrieved from http: //www.g-fras. org/en/download. html